PRAISE FOR *THE MOLECULE OF MORE*

"Daniel Lieberman and Michael Long have pulled off an amazing feat. They have made a biography of a neurotransmitter a riveting read. Once you understand the power and peril of dopamine, you'll better understand the human condition itself."
—Daniel H. Pink, author of *Drive* and *When*

"Meet a molecule whose fingerprint rests upon every aspect of human nature—from desire and drugs to politics and progress. Lieberman and Long tell the epic saga of dopamine as a page-turner that you simply can't put down."
—David Eagleman, PhD, neuroscientist at Stanford and *New York Times* bestselling author

"I've worked as an artist for forty years, and the question 'Why am I like this?' has been a puzzle, a mystery, a plea, and an occasional cry to the heavens. Lieberman and Long have created a road map for all those wrestling between insatiable longing and the here and now."
—Thomas F. Wilson, actor and comedian

"Why do we crave what we don't have rather than feel good about what we do—and why do fools fall in love? Haunting questions of human biology are answered by *The Molecule of More,* a must-read about the human condition."
—Gregg Easterbrook, author of *It's Better Than It Looks*

"As a guy who creates musical stuff for a living and reads science books for kicks, I was doubly hooked by *The Molecule of More*. Lieberman and Long lay out the astoundingly wide-ranging effects of dopamine with nimble metaphors and fat-free sentences. And the research linking creativity and madness, with dopamine as the hidden culprit—let's just say it hit home. Reading each chapter, I felt myself fitting a key smoothly into a locked door, opening onto a fresh-yet-familiar room."
—Robbie Fulks, Grammy-nominated recording artist

"Jim Watson, who deciphered the genetic code, famously said, 'There are only molecules; the rest is sociology,' adding fuel to C. P. Snow's complaint that Science and the humanities are two fundamentally different "cultures" which will never meet. The authors argue provocatively, yet convincingly, that the molecule that allows us to bridge the chasm between them is dopamine. Though written for ordinary people, the narrative is sprinkled throughout with dazzling new insights that will appeal equally to specialists."

—V.S. Ramachandran, PhD, professor at the University of California, San Diego, and at Salk Institute and author of *The Emerging Mind*

THE MOLECULE

OF MORE

THE MOLECULE
OF MORE

How a Single Chemical in Your Brain Drives
Love, Sex, and Creativity—and Will Determine
the Fate of the Human Race

DANIEL Z. LIEBERMAN, MD
AND MICHAEL E. LONG

BenBella Books, Inc.
Dallas, TX

BenBella Books, Inc.
10440 N. Central Expressway, Suite 800
Dallas, TX 75231
www.benbellabooks.com
Send feedback to feedback@benbellabooks.com

BenBella is a federally registered trademark.

Printed in the United States of America
20 19 18 17 16 15 14 13 12

ISBN 9781948836586 (trade paper)
ISBN 9781946885111 (trade cloth)
ISBN 9781946885296 (electronic)

The Library of Congress Cataloging in Publication Control Number: 2018010836

Editing by Vince Hyman
Copyediting by James Fraleigh
Proofreading by Lisa Story and Greg Teague
Indexing by Debra Bowman
Text design and composition by Aaron Edmiston
Cover design by Pete Garceau
Cover images © iStock/bauhaus1000 (lowers), nicoolay (rose), cosveta (bees)
Printed by Lake Book Manufacturing

For Sam and Zach,
who open my eyes to seeing the world in new ways.
—DZL

For Dad,
who would have told everyone even if they didn't want to hear it; and

For Kent,
who left just when things were getting interesting.
—ML

CONTENTS

CONTENTS

ACKNOWLEDGMENTS

We are most grateful to Dr. Fred H. Previc for his book *The Dopaminergic Mind in Human Evolution and History*. The book introduced us to the fundamental distinction between the future focus of dopamine and the present focus of a group of other neurotransmitters. It's written primarily for scientists, but if you are interested in a deeper look at the neurobiology that informs this book, we highly recommend it.

Thanks to our agents, Andrea Somberg and Wendy Levinson of the Harvey Klinger Agency, who immediately understood what we were doing and gave us the validation we had hoped to find. Thanks too to our publisher, Glenn Yeffeth at BenBella, whose enthusiasm and expertise put us further at ease. Thanks also to the BenBella team, especially Leah Wilson, Adrienne Lang, Jennifer Canzoneri, Alexa Stevenson, Sarah Avinger, Heather Butterfield, and everyone else there who labored over our work, even if we never met. Plus: special thanks to copy editor extraordinaire James M. Fraleigh. He could improve even this sentence, and probably in his sleep.

Dan wishes to thank Dr. Frederick Goodwin for his many years of mentorship. Dr. Goodwin is one of the world's foremost experts on bipolar disorder. He drew my attention to the relationship between immigration and bipolar genes, and also suggested that I look to Tocqueville's classic book *Democracy in America* to better understand the character of the United States in the nineteenth century. Thanks to the George Washington University Medical Faculty Associates for the opportunity to practice psychiatry in a vibrant academic environment and the privilege of treating people living with mental illness. My patients' willingness to share with me their suffering, triumphs, hopes, and fears is a constant source of inspiration for which I'm grateful. Thanks also to the medical students and trainees who ask annoyingly difficult questions, forcing me to constantly rethink my understanding of how the brain works.

Mike wishes to thank early readers Greg Northcutt and Jim & Ellen Hubbard, who confirmed that we had made the science compelling. Thanks to John J. Miller for the professional example and Peter Nash for the personal inspiration. Thanks also to my students at Georgetown University, who remind me that most of writing is thinking. I wouldn't know how to tell a story if it weren't for the late Blake Snyder, and I wouldn't know how to make it sing without Vince Gilligan—thank you, gentlemen. Thanks also to my brother Todd for the daily jest. Keep it coming. Oh yeah: Thanks, Mom.

Dan wishes to thank his wife, Masami, for her support, optimism, and good cheer. When the bumps along the road to finishing this book made me doubt myself, those doubts disappeared the moment I brought them to her. Thanks to my sons, Sam and Zach, who bring joy into my life and force me to grow as a person.

Michael wishes to thank his wife, Julia, for the last couple years of extra latitude. You always let me rant, then kiss me on the forehead and tell me I can do it anyway. Thanks also to my kids, Sam, Madeline, and Brynne, for acting interested even when you were not. Love you all.

The authors together wish to acknowledge their gratitude for the TGI Fridays near the White House where we so often indulged both control and desire dopamine. The planning and imagining that took place there would ultimately collapse into the bit of reality you now hold in your hands.

ACKNOWLEDGMENTS

Finally, this book began as an effort by two friends so uninterested in normal pastimes like fishing and baseball that the only thing we could do together was eat lunch more often or write a book. We remain friends, though a couple times it was a close call.

Daniel Z. Lieberman & Michael E. Long
February 2018

In the beginning God created the heavens and the earth.

Introduction

UP VERSUS DOWN

L ook down. What do you see? Your hands, your desk, the floor, maybe a cup of coffee, or a laptop computer or a newspaper. What do they have in common? These are things you can touch. What you see when you look down are things within your reach, things you can control right now, things you can move and manipulate with no planning, effort, or thought. Whether it's a result of your work, the kindness of others, or simple good fortune, much of what you see when you look down is yours. They're things in your possession.

Now look up. What do you see? The ceiling, perhaps pictures on a wall, or things out the window: trees, houses, buildings, clouds in the sky—whatever is in the distance. What do they have in common? To reach them, you have to plan, think, calculate. Even if it's only a little, it still requires some coordinated effort. Unlike what we see when we look down, the realm of *up* shows us things that we have to think about and work for in order to get.

Sounds simple because it is. Yet to the brain this distinction is the gateway between two wildly different ways of thinking—two utterly different ways of dealing with the world. In your brain the *down* world is managed by a handful of chemicals—neurotransmitters, they're called—that let you experience satisfaction and enjoy whatever you have in the here and now. But when you turn your attention to the world of *up*, your brain relies on a different chemical—a single molecule—that not only allows you to move beyond the realm of what's at your fingertips, but also motivates you to pursue, to control, and to possess the world beyond your immediate grasp. It drives you to seek out those things far away, both physical things and things you cannot see, such as knowledge, love, and power. Whether it's reaching across the table for the salt shaker, flying to the moon in a spaceship, or worshipping a god beyond space and time, this chemical gives us command over every distance, whether geographical or intellectual.

Those *down* chemicals—call them the *Here & Nows*—allow you to experience what's in front of you. They enable you to savor and enjoy, or perhaps to fight or run away, right now. The *up* chemical is different. It makes you desire what you don't yet have, and drives you to seek new things. It rewards you when you obey it, and makes you suffer when you don't. It is the source of creativity and, further along the spectrum, madness; it is the key to addiction and the path to recovery; it is the bit of biology that makes an ambitious executive sacrifice everything in pursuit of success, that makes successful actors and entrepreneurs and artists keep working long after they have all the money and fame they ever dreamed of; and that makes a satisfied husband or wife risk everything for the thrill of someone else. It is the source of the undeniable itch that drives scientists to find explanations and philosophers to find order, reason, and meaning.

It is why we look into the sky for redemption and God; it is why heaven is above and earth is below. It is fuel for the motor of our dreams; it is the source of our despair when we fail. It is why we seek and succeed; it is why we discover and prosper.

It is also why we are never happy for very long.

To your brain, this single molecule is the ultimate multipurpose device, urging us, through thousands of neurochemical processes, to move beyond the pleasure of just being, into exploring the universe of possibilities that come when we imagine. Mammals, reptiles, birds, and fish all have this chemical inside their brains, but no creature has more of it than a human being. It is a blessing and a curse, a motivation and a reward. Carbon, hydrogen, oxygen, plus a single nitrogen atom—it is simple in form and complex in result. This is dopamine, and it narrates no less than the story of human behavior.

And if you want to feel it right now, if you want to put it in charge, you can do that.

Look up.

A NOTE FROM THE AUTHORS

We've packed this book full of the most interesting scientific experiments we could find. Still, some parts are speculative, especially in later chapters. In addition, there are places where we oversimplify to make the material easier to understand. The brain is so complex that even the most sophisticated neuroscientist must simplify to build a model of the brain that's capable of being understood. Also, science is messy. Sometimes studies contradict one another, and it takes time to sort out which results are correct. Reviewing the entire body of evidence would quickly become tedious for the reader, so we selected studies that have influenced the field in important ways and that reflect scientific consensus, when consensus exists.

Science is not only messy; it can sometimes be bizarre. The search for understanding human behavior can take strange forms. It's not like studying chemicals in a test tube or even infections in living people. Brain researchers have to find ways to trigger important behaviors in a laboratory environment—sometimes sensitive behaviors driven by passions such as fear, greed, or sexual desire. When possible we chose studies that highlight this strangeness.

Human research in all its forms is tricky. It's not the same as clinical care, in which a doctor and a patient work together to treat the patient's illness. In that case, they choose whatever treatment they think will work best, and the only goal is to make the patient better.

The goal of research, on the other hand, is to answer a scientific question. Even though scientists work hard to minimize the risks to their participants, the science must come first. Sometimes, access to experimental treatments can be lifesaving, but usually research participants are exposed to risks they wouldn't experience in the course of regular clinical care.

By volunteering to take part in studies, participants sacrifice some of their own safety for the benefit of others—sick people who will enjoy a better life if the research is successful. It's like a firefighter running into a burning building to rescue the people trapped inside, choosing to place himself in danger for the welfare of others.

The key element, of course, is that the research participant needs to know exactly what she's getting herself into. It's called informed consent, and usually comes in the form of a lengthy document that explains the purpose of the research and lists the risks of becoming involved. It's a good system, though not perfect. Participants don't always read it carefully, especially if it's very long. Sometimes researchers leave things out because deception is an essential part of the study. But, in general, scientists do their best to make sure their participants are willing partners as they tackle the mysteries of human behavior.

Love is a need, a craving, a drive to seek life's greatest prize.
—Helen Fisher, biological anthropologist

Chapter 1

LOVE

You've found the one you waited for all your life,
so why doesn't the honeymoon last forever?

*In which we explore the chemicals that make you want sex and
fall in love—and why, sooner or later, everything changes.*

Shawn wiped a clear space on his steamed-over bathroom mirror, ran his fingers through his black hair, smiled. "This'll work," he said.

He dropped his towel and admired his flat belly. His obsession with the gym had produced two-thirds of a six-pack. From that, his mind went to a more pressing obsession: he had not been out with anyone since February. Which was a nice way of saying he hadn't had sex in seven months and three days—and he was disturbed to realize he had kept track so precisely. That streak ends tonight, he thought.

At the bar, he surveyed the possibilities. There were a lot of attractive women here tonight—not that looks were everything. He missed sex, sure, but he also missed having someone in his life, someone to text for no reason, someone who could be a welcome part of every day. He considered himself a romantic, even if tonight was just about sex.

He kept meeting the eyes of a young woman standing with a chatty friend at a high-top table. She had dark hair and brown eyes, and he noticed her because she wasn't in the usual Saturday-night uniform; she had on flats instead of heels, and she wore Levis instead of club clothes. He introduced himself and the conversation came quickly and easily. Her name was Samantha, and the first thing she said was that she was more comfortable doing cardio than putting back beers. That led to an in-depth discussion of local gyms, fitness apps, and the relative merits of working out in the morning versus the afternoon. For the rest of the night he didn't leave her side, and she quickly came to like having him there.

Lots of factors pushed them along to what would become a long-term relationship: their common interests, the ease they felt with each other, even the drinks and a little desperation. But none of that was the real key to love. The big factor was this: they were both under the influence of a mind-altering chemical. So was everyone else in the bar.

And, it turns out, so are you.

WHAT IS MORE POWERFUL THAN PLEASURE?

Dopamine was discovered in the brain in 1957 by Kathleen Montagu, a researcher working in a laboratory at the Runwell Hospital near London. Initially, dopamine was seen simply as a way for the body to produce a chemical called norepinephrine, which is what adrenaline is called when it is found in the brain. But then scientists began to observe strange things. Only 0.0005 percent of brain cells produce dopamine—one in two million—yet these cells appeared to exert an outsized influence on behavior. Research participants experienced feelings of pleasure when they turned dopamine on, and went to great lengths to trigger the activation of these rare cells. In fact, under the right circumstances, pursuit of *feel-good* dopamine activation became impossible to resist. Some scientists christened dopamine *the pleasure molecule,* and the pathway that dopamine-producing cells take through the brain was named the *reward circuit.*

The reputation of dopamine as the pleasure molecule was further cemented through experiments with drug addicts. The researchers injected them with a combination of cocaine and radioactive sugar, which allowed the scientists to figure out which parts of their brains were burning the most calories. As the intravenous cocaine took effect, participants were asked to rate how high they felt. Researchers discovered that the greater the activity in the dopamine reward pathway, the greater the high. As the body cleared the cocaine from the brain, dopamine activity decreased, and the high faded. Additional studies produced similar results. The role of dopamine as the pleasure molecule was established.

Other researchers tried to duplicate the results, and that's when unexpected things began to happen. They reasoned that it's unlikely that dopamine pathways evolved to encourage people to get high on drugs. Drugs were probably causing an artificial form of dopamine stimulation. It seemed more likely that the evolutionary processes that harnessed dopamine were driven by the need to motivate survival and reproductive activity. So they replaced cocaine with food, expecting to see the same effect. What they found surprised everyone. It was the beginning of the end for dopamine as the pleasure molecule.

Dopamine, they discovered, isn't about pleasure at all. Dopamine delivers a feeling much more influential. Understanding dopamine turns out to be the key to explaining and even *predicting* behavior across a spectacular range of human endeavors: creating art, literature, and music; seeking success; discovering new worlds and new laws of nature; thinking about God—and falling in love.

Shawn knew he was in love. His insecurities melted away. Every day made him feel on the brink of a golden future. As he spent more time with Samantha, his excitement about her grew, and his sense of anticipation became constant. Every thought of her suggested limitless possibilities. As for sex,

his libido was stronger than ever, but only for her. Other women ceased to exist. Even better, when he tried to confess all this happiness to Samantha, she interrupted him to say she felt exactly the same.

Shawn wanted to be sure they would be together forever, so one day he proposed to her. She said yes.

A few months after their honeymoon, things began to change. At the start they had been obsessed with one another, but, with the passage of time, that desperate longing became less desperate. The belief that anything was possible became less certain, less obsessive, less at the center of everything. Their elation receded. They weren't unhappy, but the profound satisfaction from their earlier time together was slipping away. The sense of limitless possibilities began to seem unrealistic. Thoughts about each other, that used to come constantly, didn't. Other women began to draw Shawn's attention, not that he intended to cheat. Samantha let herself flirt sometimes, too, even if it was no more than a shared smile with the college boy bagging groceries in the checkout line.

They were happy together, but the early gloss of their new life began to feel like their old life apart. The magic, whatever it was, was fading.

Just like my last relationship, *thought Samantha.*

Been there, done that, *thought Shawn.*

MONKEYS AND RATS AND WHY LOVE FADES

In some ways rats are easier to study than human beings. Scientists can do a lot more to them without having to worry about the research ethics board knocking at their door. To test the hypothesis that both food and drugs stimulate dopamine, the scientists implanted electrodes directly into rats' brains so they could directly measure the activity of individual dopamine neurons. Next, they built cages with chutes for food pellets. The results were just as they expected. As soon as they dropped the first pellet, the rats' dopamine systems lit up. Success! Natural rewards stimulate dopamine activity just as well as cocaine and other drugs.

Next they did something the original experimenters had not. They kept going, monitoring the rats' brains as pellets of food were dropped down the chute, day after day. The results were wholly unexpected. The rats devoured the food as enthusiastically as ever. They were obviously enjoying it. But their dopamine activity shut down. Why would dopamine stop firing when stimulation keeps coming? The answer came from an unlikely source: a monkey and a light bulb.

Wolfram Schultz is among the most influential pioneers of dopamine experimentation. As a professor of neurophysiology at the University of Fribourg, Switzerland, he became interested in the role of dopamine in learning. He implanted tiny electrodes into the brains of macaque monkeys where dopamine cells clustered together. He then placed the monkeys in an apparatus that had two lights and two boxes. Every once in a while one of the lights turned on. One light was a signal that the food pellet could be found in the box on the right. The other meant the food pellet was in the box on the left.

It took the monkeys some time to figure out the rule. At first they opened the boxes randomly, and got it right about half the time. When they found a food pellet, the dopamine cells in their brain fired, just as in the rats. After a while, the monkeys figured out the signals and reached for the correct, food-containing box every time—and at that, the timing of the dopamine release began to change from firing at the discovery of the food to firing at the light. Why?

Seeing the light go on would always be unexpected. But once the monkeys figured out that the light meant they were about to get food, the "surprise" they felt came exclusively from the appearance of the light, not from the food. From that, a new hypothesis arose: dopamine activity is not a marker of pleasure. It is a reaction to the unexpected—to possibility and anticipation.

As human beings, we get a dopamine rush from similar, promising surprises: the arrival of a sweet note from your lover (*What will it say?*), an email message from a friend you haven't seen in years (*What's the news going to be?*), or, if you're looking for romance, meeting a fascinating new partner at a sticky table in the same old bar (*What might happen?*). But when these things become regular events, their novelty fades, and so does the dopamine rush—and a sweeter note or a longer email or a better table won't bring it back.

This simple idea provides a chemical explanation for an age-old question: Why does love fade? Our brains are programmed to crave the unexpected and thus to look to the future, where every exciting possibility begins. But when anything, including love, becomes familiar, that excitement slips away, and new things draw our attention.

The scientists who studied this phenomenon named the buzz we get from novelty *reward prediction error*, and it means just what the name says. We constantly make predictions about what's coming next, from what time we can leave work, to how much money we expect to find when we check our balance at the ATM. When what happens is better than what we expect, it is literally an error in our forecast of the future: Maybe we get to leave work early, or we find a hundred dollars more in checking than we expected. That happy error is what launches dopamine into action. It's not the extra time or the extra money themselves. It's the thrill of the unexpected good news.

In fact, the mere possibility of a reward prediction error is enough for dopamine to swing into action. Imagine you're walking to work on a familiar street, one you've traveled many times before. All of a sudden you notice that a new bakery has opened, one you've never seen. You immediately want to go in and see what they have. That's dopamine taking charge, and it produces a feeling different from enjoying how something tastes, feels, or looks. It's the pleasure of anticipation—the possibility of something unfamiliar and better. You're excited about the bakery, yet you haven't eaten any of their pastries, sampled any of their coffee, or even seen how it looks inside.

You go in and order a cup of dark roast and a croissant. You take a sip of the coffee. The complex flavors play across your tongue. It's the best you've ever had. Next you take a bite of the croissant. It's buttery and flaky, exactly like the one you had years ago at a café in Paris. Now how do you feel? Maybe that your life is a little better with this new way to start your day. From now on you're going to come here every morning for breakfast, and have the best coffee and flakiest croissant in the city. You'll tell your friends about it, probably more than they care to hear. You'll buy a mug with the café's name on it. You'll even be more excited to start the day because, *well, this awesome café, that's why.* That's dopamine in action.

It's as if you have fallen in love with the café.

Yet sometimes when we get the things we want, it's not as pleasant as we expect. Dopaminergic excitement (that is, the thrill of anticipation) doesn't last forever, because eventually the future becomes the present. The thrilling mystery of the unknown becomes the boring familiarity of the everyday, at which point dopamine's job is done, and the letdown sets in. The coffee and croissants were so good, you made that bakery your regular breakfast stop. But after a few weeks, "the best coffee and croissant in the city" became the same old breakfast.

But it wasn't the coffee and the croissant that changed; it was your expectation.

In the same way, Samantha and Shawn were obsessed with each other until their relationship became utterly familiar. When things become part of the daily routine, there is no more reward prediction error, and dopamine is no longer triggered to give you those feelings of excitement. Shawn and Samantha surprised each other in a sea of anonymous faces at a bar, then obsessed over each other until the imagined future of never-ending delight became the concrete experience of reality. Dopamine's job—and ability—to idealize the unknown came to an end, so dopamine shut down.

Passion rises when we dream of a world of possibility, and fades when we are confronted by reality. When the god or goddess of love beckoning you to the boudoir becomes a sleepy spouse blowing his or her nose into a ratty Kleenex, the nature of love—the reason to stay—must change from dopaminergic dreams to . . . something else. But what?

ONE BRAIN, TWO WORLDS

John Douglas Pettigrew, emeritus professor of physiology at the University of Queensland, Australia, is a native of the delightfully named city of Wagga Wagga. Pettigrew had a brilliant career as a neuroscientist, and is best known for updating the flying primates theory, which established bats as our distant cousins. While working on this idea, Pettigrew became the first person to clarify how the brain creates a three-dimensional map of the world. That sounds far removed from passionate relationships, but it would turn out to be a key concept for explaining dopamine and love.

Pettigrew found that the brain manages the external world by dividing it into separate regions, the *peripersonal* and the *extrapersonal*—basically, near and far. Peripersonal space includes whatever is in arm's reach; things you can control right now by using your hands. This is the world of what's real, right now. Extrapersonal space refers to everything else—whatever you can't touch unless you move beyond your arm's reach, whether it's three feet or three million miles away. This is the realm of possibility.

With those definitions in place, another fact follows, obvious but useful: since moving from one place to another takes time, any interaction in the extrapersonal space must occur in the future. Or, to put it another way, distance is linked to time. For instance, if you're in the mood for a peach, but the closest one is sitting in a bin at the corner market, you can't enjoy it now. You can only enjoy it in the future, after you go get it. Acquiring something out of your reach may also take some planning. It could be as simple as standing up to turn on a light, walking to the market for that peach, or figuring out how to launch a rocket to get to the moon. This is the defining characteristic of things in the extrapersonal space: to get them requires effort, time, and in many cases, planning. By contrast, anything in the peripersonal space can be experienced in the here and now. Those experiences are immediate. We touch, taste, hold, and squeeze; we feel happiness, sadness, anger, and joy.

This brings us to a clarifying fact of neurochemistry: the brain works one way in the peripersonal space and another way in the extrapersonal space. If you were designing the human mind, it makes sense that you would create a brain that distinguishes between things in this way, one system for what you have and another for what you don't. For early humans, the familiar phrase "either you have it or you don't" could be translated into "either you have it or you're dead."

From an evolutionary standpoint, food that you don't have is critically different from food that you do have. It's the same for water, shelter, and tools. The division is so fundamental that separate pathways and chemicals evolved in the brain to handle peripersonal and extrapersonal space. When you look down, you look into the peripersonal space, and for that the brain is controlled by a host of chemicals concerned with experience in the here and now. But when the brain is engaged with the extrapersonal space, one chemical exercises more control than all the others, the chemical associated with anticipation and possibility: dopamine. Things in the distance, things we don't have yet, cannot be used or consumed, only desired. Dopamine has a very specific job: maximizing resources that will be available to us in the future; the pursuit of better things.

Every part of living is divided in this way: we have one way of dealing with what we want, and another way of dealing with what we have. Wanting a house, experiencing the kind of desire that motivates the hard work necessary to find it and purchase it, uses a different set of brain circuits than enjoying it once it's yours. Anticipating a raise activates future-oriented dopamine, and it feels very different from the here-and-now experience of receiving the larger paycheck for the second or third time. And finding love takes a different set of skills than making love stay. Love must shift from an extrapersonal experience to a peripersonal one—from pursuit to possession; from something we anticipate to something we have to take care of. These are vastly different skills, which is why over time the nature of love has to change—and why, for so many people, love fades away at the end of the dopamine thrill we call romance.

Yet many people make the transition. How do they do it—how are they outsmarting the seduction of dopamine?

🦋 GLAMOUR 🦋

Glamour is a beautiful illusion—the word "glamour" originally meant a literal magic spell—that promises to transcend ordinary life and make the ideal real. It depends on a special combination of mystery and grace. Too much information breaks the spell.

—Virginia Postrel

Glamour is present when we see things that stimulate our dopaminergic imagination, drowning out our ability to accurately perceive here-and-now reality.

A good example is air travel. Look up. Is there an airplane in the sky? What kinds of thoughts and feelings are triggered? Many people experience a longing to be on the plane, traveling to exotic locations that are far away—a carefree getaway that begins with a ride among the clouds. Of course, if you were on the plane, your here-and-now senses would inform you that this paradise in the sky is more like a rush-hour bus across town: cramped, exhausting, and unpleasant—the opposite of elegant.

Likewise, what could be more glamorous than Hollywood? Beautiful actors and actresses go to parties, stand around swimming pools, and flirt. The reality is far different, involving 14-hour days sweating under hot lights. Women actors are exploited sexually and men are pressured to take steroids and growth hormone to get the fabulous bodies we see on screen. Gwyneth Paltrow, Megan Fox, Charlize Theron, and Marilyn Monroe have all described "casting couch" experiences (all but Marilyn Monroe said they declined the offer to trade sex for a coveted role). Nick Nolte, Charlie Sheen, Mickey Rourke, and Arnold Schwarzenegger have all admitted to using steroids, which can cause liver damage, mood swings, violent outbursts, and psychosis. It's a tawdry business.

Mountains aren't tawdry, though. They're majestic, standing far off in the distance, softened by the blurring effect of miles of air, like a soft-focus photograph of a bride on her wedding day. Those with higher levels of dopamine want to climb it, explore it, conquer it. But they can't, because it doesn't exist. The mountain itself exists. But the imagined experience of being on it is impossible to achieve. The reality is that most of the time you're on a mountain you can't even tell. Typically you're surrounded by trees, and that's all you see. Occasionally you might come to a scenic overlook in which you can see for miles over the valley. But as you look, it's the far-away valley that's full of promise and beauty, not the mountain you're standing on. Glamour creates desires that cannot be fulfilled because they are desires for things that exist only in the imagination.

Whether it's an airplane in the sky, a movie star in Hollywood, or a distant mountain, only things that are out of reach can be glamorous; only things that are unreal. Glamour is a lie.

One day at lunch, Samantha ran into Demarco, her last serious boyfriend before Shawn. They hadn't seen each other in years, hadn't even come across each other on Facebook. She found him as funny and smart as ever, and in great shape, too. In minutes she was starry-eyed again. Here was something she hadn't felt in a long time, the surge of excitement and the sense of possibility with a man who connected with her, someone who seemed full of fresh things for her to discover. He was excited, too, and anxious to share his feelings. The first thing he shared was how excited he was to be engaged. His fiancée was "the one" and he hoped Samantha would meet her, because he had never cared for someone so special as this new woman.

After Demarco left, Samantha decided this was a good day to drink. She adjourned to the bar and ordered a basket of tortilla chips and a Miller Lite, and spent the next half-hour picking at the label. She loved Shawn, she really did—or did she? They had been in a rut for most of a year. That feeling with Demarco was what she wanted. She had once had it with Shawn, but not anymore.

THE DARK SIDE

There's a dark side to dopamine. If you drop a pellet of food into a rat's cage, the animal will experience a dopamine surge. Who knew that the world was a place where food dropped from the sky? But if you keep dropping pellets every 5 minutes, dopamine stops. The rat knows when to expect the food, so there's no surprise, and there is no *error* in the rat's *prediction* of a *reward*. But what if you drop the pellet at random times, so it's always a surprise? And what if, instead of rats and food pellets, you replace them with people and money?

Picture the busy floor of a casino with a crowded blackjack table, a high-stakes poker game, and a spinning roulette wheel. It's the epitome of Vegas glitz, but casino operators know that these high-roller games are not where the biggest profits are made. Those come from the lowly slot machine, beloved by tourists, retirees, and workaday gamblers who drop in daily for a few hours alone with flashing lights, ringing bells, and clicking wheels. The modern standard for casino design is to dedicate a whopping 80 percent of floor space to slot machines, and for good reason: slot machines bring in the majority of casino gambling revenue.

One of the world's largest manufacturers of slot machines is owned by a company called Scientific Games. Science plays a big role in the design of these compelling devices. Although slot machines date back to the nineteenth century, modern refinements are based on the pioneering work of behavioral scientist B. F. Skinner, who in the 1960s mapped out the principles of behavior manipulation.

In one experiment Skinner placed a pigeon in a box. He found that he could condition it to peck a lever to get a pellet of food. Some experiments used one peck, others ten, but the number required never changed within any single experiment. The results weren't particularly interesting. Regardless of the number of presses required, each pigeon pecked at its lever like a bureaucrat stamping an endless pile of documents.

Then Skinner tried something different. He set up an experiment in which the number of presses needed to release a pellet changed randomly. Now the pigeon never knew when the food would come. Every reward was unexpected. The birds became excited. They pecked faster. Something was spurring them on to greater efforts. Dopamine, the molecule of surprise, had been harnessed, and the scientific foundation of the slot machine was born.

When Samantha saw her old boyfriend, all the feelings came rushing back—excitement, possibility, focus, butterflies. She wasn't on the prowl for romance, but she didn't have to be. Demarco's appearance, and the half-conscious dream of another chance at passionate excitement, was an unexpected treat dropped into her emotional life, and that surprise was the source of her excitement. Samantha, of course, didn't know that.

She and Demarco decide to meet again for a drink, and it goes well. They decide to have lunch the next day, too, and pretty soon their meetings become a standing "date." The feelings are exhilarating. They touch when they talk. They hug when they part. When they are together, the time flies, just like when they dated before—and, when she thinks about it, just like it used to be with Shawn. *Maybe*, she thinks, *Demarco's the one.* But with an understanding of the role of dopamine, it's clear that this relationship is not something new. It's just another repetition of dopamine-driven excitement.

The novelty that triggers dopamine doesn't go on forever. When it comes to love, the loss of passionate romance will always happen eventually, and then comes a choice. We can transition to a love that's fed by a day-to-day appreciation of that other person in the here and now, or we can end the relationship and go in search of another roller coaster ride. Choosing the dopaminergic kick takes little effort, but it ends fast, like the pleasure of eating a Twinkie. Love that lasts shifts the emphasis from anticipation to experience; from the fantasy of anything being possible to engagement with reality and all its imperfections. The transition is difficult, and when the world presents an easy way out of a difficult task, we tend to take it. That's why, when the dopamine firing of early romance ends, many relationships end, too.

Early love is a ride on a merry-go-round that sits at the foot of a bridge. That carousel can take you around and around on a beautiful trip as many times as you like, but it will always leave you where you began. Each time the music stops and your feet are back on the ground, you must make a choice: take one more whirl, or cross that bridge to another, more enduring kind of love.

MICK JAGGER, GEORGE COSTANZA, AND "SATISFACTION"

When Mick Jagger first sang "I can't get no satisfaction!" in 1965, we could not have known that he was predicting the future. As Jagger told his biographer in 2013, he has been with about four thousand women—a different partner every ten days of his adult life.

Note that Mick didn't follow up with, ". . . and at four thousand, I finally found satisfaction. I'm done!" Presumably he'll keep going as long as he can. So how many lovers would be enough to get "satisfaction"? If you've had four thousand, we can safely say that dopamine is steering things in your life, at least when it comes to sex. And dopamine's prime directive is *more*. If Sir Mick chases satisfaction another half century, he still won't catch it. His idea of satisfaction is not satisfaction at all. It's pursuit, which is driven by dopamine, the molecule that cultivates perpetual dissatisfaction. After he beds a lover, his immediate goal will be to find another.

In this way, Mick isn't alone. He isn't even unusual. Mick Jagger is just a confident version of TV's George Costanza. In nearly every episode of *Seinfeld*, George fell in love. He went to ridiculous lengths to get a date, and he was capable of almost anything if it might lead to sex. He imagined each new woman as a potential life mate, the perfect female who would go with him into happily ever after. But every *Seinfeld* fan knows how those stories ended. George would be crazy about the woman up until the moment she returned his affection. When he didn't have to try anymore, all he wanted was out. George Louis Costanza was so addicted to the dopamine thrill of chasing romance that he spent an entire season trying to extract himself from his engagement to the only woman who continued to love him despite every awful thing he did. And when his fiancée died from licking toxic glue on the envelopes of their wedding invitations, George wasn't devastated. He was relieved, even joyful. He was ecstatic to rejoin the chase. Mick is like George, and George is like all of us. We revel in the passion, the focus, the excitement, the thrill of finding new love. The difference is that most of us figure out at some point that dopamine lies to us. Unlike the former latex salesman for Vandelay Industries and the lead singer of the Rolling Stones, we come to understand that the next beautiful woman or a handsome man we see is probably not the key to "satisfaction."

"How's Shawn?" said Samantha's mother.

"Well . . . ," Samantha traced the rim of her coffee cup. "This isn't the way I expected it to be."

"Again?"

"Here it comes," said Samantha.

"I'm just saying that Shawn seems like a great guy—"

"Mother, I don't want to play 'count your blessings.' "

"This isn't the first time. Remember Lawrence? And Demarco?" Samantha bit her lip. "Why can't you enjoy the things you have?"

THE CHEMICAL KEYS TO LONG-LASTING LOVE

From dopamine's point of view, having things is uninteresting. It's only getting things that matters. If you live under a bridge, dopamine makes you want a tent. If you live in a tent, dopamine makes you want a house. If you live in the most expensive mansion in the world, dopamine makes you want a castle on the moon. Dopamine has no standard for good, and seeks no finish line. The dopamine circuits in the brain can be stimulated only by the possibility of whatever is shiny and new, never mind how perfect things are at the moment. The dopamine motto is "More."

Dopamine is one of the instigators of love, the source of the spark that sets off all that follows. But for love to continue beyond that stage, the nature of the love relationship has to change because the chemical symphony behind it changes. Dopamine isn't the pleasure molecule, after all. It's the anticipation molecule. To enjoy the things we have, as opposed to the things that are only possible, our brains must transition from future-oriented dopamine to present-oriented chemicals, a collection of neurotransmitters we call the *Here and Now* molecules, or the H&Ns. Most people have heard of the H&Ns. They include serotonin, oxytocin, endorphins (your brain's version of morphine), and a class of chemicals called endocannabinoids (your brain's version of marijuana). As opposed to the pleasure of anticipation via dopamine, these chemicals give us pleasure from sensation and emotion. In fact, one of the endocannabinoid molecules is called anandamide, named after a Sanskrit word that means *joy, bliss, and delight.*

According to anthropologist Helen Fisher, early or "passionate" love lasts only twelve to eighteen months. After that, for a couple to remain attached to one another, they need to develop a different sort of love called *companionate* love. Companionate love is mediated by the H&Ns because it involves experiences that are happening right here, right now—*you're with the one you love, so enjoy it.*

Companionate love is not a uniquely human phenomenon. We see it among animal species that mate for life. Their behavior is characterized by cooperative territory defense and nest building. The bonded pair feed each other, groom each other, and share parental chores. Most of all, they stay close to each other and display expressions of anxiety when separated. It's the same for humans. Humans engage in similar activities and have similar feelings, particularly satisfaction that there is another person whose life is deeply entwined with their own.

When the H&Ns take over in the second stage of love, dopamine is suppressed. It has to be because dopamine paints a picture in our minds of a rosy future in order to spur us on through the hard work necessary to make it a reality. Dissatisfaction with the present state of affairs is an important ingredient in bringing about change, which is what a new relationship is all about. H&N companionate love, on the other hand, is characterized by deep and enduring satisfaction with the present reality, and an aversion to change, at least with regard to one's relationship with one's partner. In fact, though dopamine and H&N circuits can work together, under most circumstances they counter each other. When H&N circuits are activated, we are prompted to experience the real world around us, and dopamine is suppressed; when dopamine circuits are activated, we move into a future of possibilities, and H&Ns are suppressed.

Laboratory testing supports this idea. When scientists looked at blood cells extracted from people who were in the passionate stage of love, they found lower levels of H&N serotonin receptors compared to "healthy" people, an indicator that the H&Ns were in retreat.

It's not easy to say farewell to the dopaminergic thrill of new partners and passionate longing, but the ability to do so is a sign of maturity, and a step toward long-lasting happiness. Think of a man who plans a vacation to Rome. He spends weeks scheduling each day, making sure he will be able to visit all the museums and landmarks he's heard so much about. But when he stands among the most beautiful artwork ever created, he thinks about how he's going to get to the restaurant where he has reservations for dinner. He's not ungrateful to see the masterpieces of Michelangelo. It's just that his personality is primarily dopaminergic: he enjoys anticipation and planning more than doing. Lovers experience the same disconnect between anticipation and experience. The early part, *passionate love*, is dopaminergic—exhilarating, idealized, curious, future looking. The later part, *companionate love*, is H&N focused—satisfying, peaceful, and experienced through bodily senses and emotions.

A romance built on dopamine is a thrilling, if short-lived, roller coaster ride, but our brain chemistry gives us the tools to move down the path that leads to companionate love. Just as dopamine is the molecule of obsessive yearning, the chemicals most associated with long-term relationships are oxytocin and vasopressin. Oxytocin is more active in women and vasopressin in men.

Scientists have studied these neurotransmitters in the laboratory in a variety of animals. For example, when scientists injected oxytocin into the brains of female prairie voles, the animals formed a long-term bond with whatever male happened to be around. Similarly, when male voles that were genetically programmed to be promiscuous were given a gene that boosted vasopressin, they mated with one female exclusively, even though other receptive females were available. Vasopressin acted like a "good-husband hormone." Dopamine does the opposite. Human beings who have genes that produce high levels of dopamine have the highest number of sexual partners and the lowest age of first sexual intercourse.

Most couples have sex less frequently as obsessive dopaminergic love evolves into companionate H&N love. This makes sense, since oxytocin and vasopressin suppress the release of testosterone. In a similar way, testosterone suppresses the release of oxytocin and vasopressin, which helps explain why men with naturally high quantities of testosterone in their blood are less likely to marry. Similarly, single men have more testosterone than married men. And if a man's marriage becomes unstable, his vasopressin falls, and his testosterone goes up.

Do human beings require long-term companionship? There's good evidence that the answer is yes. Despite the superficial appeal of having multiple partners, most people eventually settle down. A United Nations survey found that more than 90 percent of men and women marry by the age of forty-nine. We can live without companionate love, but the majority of us arrange a good portion of our lives around trying to find it and keep it. The H&Ns give us the ability to do that. They allow us to find satisfaction in what our senses deliver—what is right in front of us, and what we can experience without the nagging sense that we need something more.

TESTOSTERONE: THE HERE & NOW CHEMICAL OF SEXUAL ATTRACTION

The night Samantha first met Shawn, she was on day thirteen of her menstrual cycle. Why does that matter?

Testosterone drives sexual desire in both men and women. Men produce large amounts—it's responsible for aspects of masculinity such as facial hair, increased muscle mass, and a low-pitched voice. Women produce smaller amounts in their ovaries. On average, women have the highest levels of testosterone on days thirteen and fourteen of their menstrual cycle. That's when the egg is released from the ovary, and they are most likely to get pregnant. There are also random variations from day to day and even within a day. Some women produce more testosterone in the morning, others later in the day. The largest variation is between individuals; some women naturally produce more than others. Testosterone can even be administered as a drug. When scientists at Procter & Gamble (the maker of Old Spice cologne and Pampers diapers) applied a testosterone gel to women's skin, the women had more sex. Unfortunately, some of the women developed facial hair, low-pitched voices, and male pattern baldness, so the "female Viagra" gel never received Food and Drug Administration (FDA) approval in the United States.

Helen Fisher, an anthropologist at Rutgers University and chief scientific advisor to the Internet dating site Match .com, points out that the type of sexual drive testosterone produces is similar to other natural urges, such as hunger.

When one is hungry, all kinds of different foods will satisfy the urge to eat. Similarly, when a person experiences testosterone-induced sexual urges, the desire is for sex in general, not necessarily for a particular person. In many cases, especially with young people, nearly anyone will do. Neither is it an overwhelming desire. People don't die from sexual hunger. Testosterone doesn't drive them to commit suicide or murder—unlike the dopaminergic experience of being overwhelmed by love.

Shawn wiped a clear space in the steamed-over bathroom mirror, ran his fingers through his black hair, smiled. "This'll work," he said.

"Wait. Hold still," said Samantha. She swept a lock off his forehead. "This'll make you look so handsome."

"And then . . ."

"Down, boy," said Samantha, and she gave him a peck on the cheek.

DOPAMINE GETS YOU INTO BED . . . AND THEN GETS IN THE WAY

From eager anticipation to the physical pleasures of intimacy, the stages of sex recapitulate the stages of love: sex is love on fast forward. Sex begins with desire, a dopaminergic phenomenon driven by the hormone testosterone. It continues with arousal, another forward-looking, dopaminergic experience. As physical contact begins, the brain shifts control to the H&Ns to deliver the pleasure of the sensory experience, mainly with the release of endorphins. The consummation of the act, orgasm, is almost entirely a here-and-now experience, with endorphins and other H&N neurotransmitters working together to shut down dopamine.

This transition was caught on camera when men and women in the Netherlands were placed in brain scanners and then stimulated to orgasm. The scans showed that sexual climax was associated with decreased activation throughout the prefrontal cortex, a dopaminergic part of the brain responsible for placing deliberate restrictions on behavior. The relaxation of control allowed the activation of H&N circuits necessary for sexual climax. It didn't matter whether the person being tested was a man or a woman. With few exceptions the brain's response to orgasm was the same: dopamine off, H&N on.

That's how it's supposed to be. But just as some people have difficulty moving from passionate love to companionate love, it can also be difficult for dopamine-driven people to let the H&Ns take over during sex. That is, highly driven women and men sometimes find it a significant challenge to turn off their thoughts and just experience the sensations of intimacy—to think less and feel more.

While H&N neurotransmitters let us experience reality—and reality during sex is intense—dopamine floats above reality. It is always able to conjure up something better. To add to its seduction, it puts us in control of that alternate reality. That these imagined worlds may be impossible doesn't matter. Dopamine can always send us chasing phantoms.

Sexual encounters, especially those within ongoing relationships, fall prey to these dopamine phantoms all the time. A survey of 141 women found that 65 percent of them daydreamed during intercourse about being with another person or even doing something completely different. Other studies have put the figure as high as 92 percent. Men daydream during sex about as much as women, and the more sex both men and women have, the more likely they are to daydream.

It is ironic that brain circuits that give us the energy and motivation we need to get ourselves into bed with a desirable partner subsequently get in the way of our enjoying the fun. Part of it may involve the intensity of the experience. Sex for the first time is more intense than sex for the hundredth time—especially sex for the hundredth time with the same partner. But the climax of the experience, orgasm, is almost always intense enough to move even the most detached dreamer into the immediate world of H&N.

WHY MOM WANTS YOU TO WAIT UNTIL YOU'RE MARRIED

Though changes in culture have made the attitude passé in some quarters, there are still a lot of mothers (and anxious fathers) who encourage their daughters to "save themselves for marriage." This is often a part of a larger moral or religious teaching, but is there any advantage to waiting that is based in brain chemistry?

Testosterone and dopamine have a special relationship. During passionate love, testosterone is the one H&N that is not suppressed in favor of dopamine. In fact, they work together to form a feedback loop—a perpetual motion machine that enhances our feelings of romance. Passionate love usually increases the desire to have sex. Testosterone revs up that desire. Increased desire in turn increases passionate love. Therefore, *denying* sexual satisfaction actually enhances passion—not necessarily forever, of course, and not without significant sacrifice, but the effect is real. Thus we find a chemical explanation that, long ago, may have been at least part of the basis for behavior we see today. Waiting prolongs the most exciting phase of love. The bittersweet feelings of distance and denial are the business end of a chemical reaction.

Passion deferred is passion sustained. If mom wants her daughter to get married, amplifying the passion is a good way to help things along. Dopamine tends to shut down once fantasy becomes reality, and dopamine is the driving chemical of romantic love. So what would raise dopamine more: agreeing to sex now, or keeping it in the future? Mom knows the answer, even if we're only now learning why.

Shawn had gained a little weight, but Samantha found him more attractive than ever. Shawn thought Samantha looked better than ever, too. Even though he appreciated how great she looked dressed up, he confided to his friends that nothing was sexier to him than when she woke up with tangled hair and no makeup, wearing one of his old T-shirts from college. Lately they kept their voices low to steal a few extra minutes while the baby slept, because the morning, alone and unguarded, was a rare moment when they could enjoy the presence of each other.

Samantha had learned how to help Shawn overcome the insecurities that held him back at work, and he found ways to free up time for her, so she could pursue her master's degree. But more and more they just savored each other's company. Sometimes they didn't talk at all, and while once that would have seemed strange to them, these days it just felt right. Samantha remembered the night Shawn reached for her, stroked her hip, then took back his hand. She heard him flip over and make the sound he always made just before going to sleep.

"What's wrong?" she asked.

"Nothing," said Shawn. "Just making sure you're there."

Dopamine got the nickname "the pleasure molecule" based on experiments with addictive drugs. The drugs lit up dopamine circuits, and test participants experienced euphoria. It seemed simple until studies done with natural rewards—food, for example—found that only unexpected rewards triggered dopamine release. Dopamine responded not to reward, but to reward prediction error: the actual reward minus the expected reward. That's why falling in love doesn't last forever. When we fall in love, we look to a future made perfect by the presence of our beloved. It's a future built on a fevered imagination that falls to pieces when reality reasserts itself twelve to eighteen months later. Then what? In many cases it's over. The relationship comes to an end, and the search for a dopaminergic thrill begins all over again. Alternatively, the passionate love can be transformed into something more enduring. It can become companionate love, which may not thrill the way dopamine does, but has the power to deliver happiness—long-term happiness based on H&N neurotransmitters such as oxytocin, vasopressin, and endorphin.

It's like our favorite old haunts—restaurants, shops, even cities. Our affection for them comes from taking pleasure in the familiar ambience: the real, physical nature of the place. We enjoy the familiar not for what it could become, but for what it is. That is the only stable basis for a long-term, satisfying relationship. Dopamine, the neurotransmitter whose purpose is to maximize future rewards, starts us down the road to love. It revs our desires, illuminates our imagination, and draws us into a relationship on an incandescent promise. But when it comes to love, dopamine is a place to begin, not to finish. It can never be satisfied. Dopamine can only say, "More."

FURTHER READING

Fowler, J. S., Volkow, N. D., Wolf, A. P., Dewey, S. L., Schlyer, D. J., MacGregor, R. R., . . . Christman, D. (1989). Mapping cocaine binding sites in human and baboon brain in vivo. *Synapse, 4*(4), 371–377.

Colombo, M. (2014). Deep and beautiful. The reward prediction error hypothesis of dopamine. *Studies in History and Philosophy of Science Part C: Studies in History and Philosophy of Biological and Biomedical Sciences, 45,* 57–67.

Previc, F. H. (1998). The neuropsychology of 3-D space. *Psychological Bulletin, 124*(2), 123.

Skinner, B. F. (1990). *The behavior of organisms: An experimental analysis.* Cambridge, MA: B. F. Skinner Foundation.

Fisher, H. E., Aron, A., & Brown, L. L. (2006). Romantic love: A mammalian brain system for mate choice. *Philosophical Transactions of the Royal Society of London B: Biological Sciences, 361*(1476), 2173–2186.

Marazziti, D., Akiskal, H. S., Rossi, A., & Cassano, G. B. (1999). Alteration of the platelet serotonin transporter in romantic love. *Psychological Medicine, 29*(3), 741–745.

Spark, R. F. (2005). Intrinsa fails to impress FDA advisory panel. *International Journal of Impotence Research, 17*(3), 283–284.

Fisher, H. (2004). *Why we love: The nature and chemistry of romantic love.* New York: Macmillan.

Stoléru, S., Fonteille, V., Cornélis, C., Joyal, C., & Moulier, V. (2012). Functional neuroimaging studies of sexual arousal and orgasm in healthy men and women: A review and meta-analysis. *Neuroscience & Biobehavioral Reviews, 36*(6), 1481–1509.

Georgiadis, J. R., Kringelbach, M. L., & Pfaus, J. G. (2012). Sex for fun: A synthesis of human and animal neurobiology. *Nature Reviews Urology, 9*(9), 486–498.

Garcia, J. R., MacKillop, J., Aller, E. L., Merriwether, A. M., Wilson, D. S., & Lum, J. K. (2010). Associations between dopamine D4 receptor gene variation with both infidelity and sexual promiscuity. *PLoS One, 5*(11), e14162.

Komisaruk, B. R., Whipple, B., Crawford, A., Grimes, S., Liu, W. C., Kalnin, A., & Mosier, K. (2004). Brain activation during vaginocervical self-stimulation and orgasm in women with complete spinal cord injury: fMRI evidence of mediation by the vagus nerves. *Brain Research, 1024*(1), 77–88.

Chapter 2
DRUGS

You want it . . . but will you like it?

*In which dopamine overwhelms reason to create consuming
desire for the most destructive behaviors imaginable.*

*A guy walks past a restaurant, smells burgers cooking. He imagines taking
a bite; he can almost taste it. He's on a diet, but at this point he can't
think of anything he wants more than that hamburger, so he goes in and
orders one. Sure enough, the first bite is wonderful, but the second bite, not
so much. With each bite, his enjoyment is less and less—so much for the
hoped-for "hamburger heaven." He finishes anyway, not really knowing
why, then feels a little nauseated and very much defeated because he didn't
stick to his diet.*

*As he heads back into the street, the thought crosses his mind: there's a
big difference between wanting something and liking it.*

WHO'S IN CONTROL OF YOUR BRAIN?

At some point, everyone asks the question, why? *Why do I do the things that I do? Why do I make the choices that I make?*

On the surface, this seems like an easy question: we do things for a reason. We put on a sweater because we're cold. We get up in the morning and go to work because we need to pay the bills. We brush our teeth to prevent tooth decay. Most of what we do is for the sake of other things; things such as feeling warm, having money to pay bills, and to avoid being scolded by the dentist.

The problem is that you can ask this question as long as you like. Why do we want to stay warm? Why do we care if we pay the bills? Why do we want to avoid the dentist's scolding? Children play this game all the time: "It's time to go to bed." Why? "Because you need to get up for school in the morning." Why? "Because you need an education." Why? And so on.

The philosopher Aristotle played this same game, but with a more serious purpose. He looked at all the things we do for the sake of something else and wondered if there was an end to it all. Why do you go to work, really? Why do you need to make money? Why do you have to pay bills? Why do you want the electricity to stay on? Where does it end? Is there anything we seek for itself only, not because it leads to something else? Aristotle decided there was. He decided there was a single thing that lay at the end of every string of *Whys*, and its name was Happiness. Everything we do, ultimately, is for the sake of happiness.

It's hard to argue with this conclusion. After all, it makes us happy to be able to pay our bills and have electricity. It makes us happy to have healthy teeth and educated minds. It may even make us happy to suffer pain, if we're doing it for a worthy cause. Happiness is the polestar that guides our journey through life. When faced with a range of options, we choose the one that leads to the most happiness.

Except we don't.

Our brains aren't wired that way. Think of how many people you know who just "fell into" their careers, or who chose their college based on nothing but a gut feeling that it was the right one. Only once in a while do we sit down to consider our options rationally, weighing one against the other. Such an exercise is tiring work, and the outcome is rarely satisfying. We seldom reach the point at which we can say with certainty that we made the right decision. It's much easier just to do what we want, so that's what we do.

The next question, of course, becomes, "Well, then, what do we want?" The answer depends on whom you ask: one person might want to be rich, another might want to be a good father. The answer depends on when you ask, too. The 7:00 PM answer might be "dinner"; the 7:00 AM answer might be "another 10 minutes of sleep." Sometimes people don't know what they want at all; other times they want lots of things at once—things that they cannot have at the same time, because they conflict with one another. Most people, when they see a donut, want to eat it. Most people, when they see a donut, want to not eat it. What's going on?

HOW TO STAY ALIVE

Andrew was a young man in his twenties who worked for a company that sold enterprise software. He had a confident, outgoing personality, and was one of the top salespeople in the company. He was so consumed with his work that he spent almost no time relaxing or pursuing other activities, except one: picking up women. He estimated that he had slept with over a hundred women but had never experienced an intimate relationship with any of them. It was something he longed for, something he knew was important for his long-term happiness, and he recognized that continuing his pattern of one-night stands wasn't going to get him there. Nevertheless, the pattern continued.

Wanting, or desire, flows from an evolutionarily old part of the brain deep inside the skull called the ventral tegmental area. It is rich in dopamine; in fact, it is one of the two main dopamine-producing regions. Like most brain cells, the cells that grow there have long tails that wind through the brain until they reach a place called the nucleus accumbens. When these long-tailed cells are activated, they release dopamine into the nucleus accumbens, driving the feeling we know as motivation. The scientific term for this circuit is the mesolimbic pathway, although it's easier to simply call it the *dopamine desire circuit (Figure 1)*.

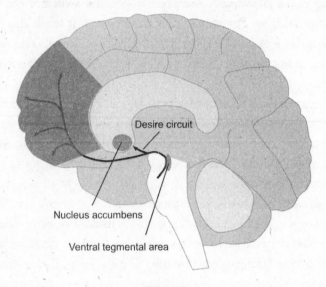

Figure 1

This dopamine circuit evolved to promote behaviors that lead to survival and reproduction, or, to put it more plainly, to help us get food and sex, and to win competitions. It's the desire circuit that's activated when you see the plate of donuts on the table, and it's activated not by need, but by the presence of something attractive from an evolutionary or life-sustaining standpoint. That is, at the moment such a thing is seen, the circuit is activated whether or not you're hungry. That's the nature of dopamine. It's always focused on acquiring more of everything with an eye toward providing for the future. Hunger is something that happens here and now, in the present. But dopamine says, "Go ahead and eat the donut, even if you're not hungry. It will increase your chance of staying alive in the future. Who knows when food will be available next?" That made sense for our evolutionary ancestors, who lived most of their lives on the brink of starvation.

For a biological organism, the most important goal related to the future is to be alive when it comes. As a result, the dopamine system is more or less obsessed with keeping us alive. It constantly scans the environment for new sources of food, shelter, mating opportunities, and other resources that will keep our DNA replicating. When it finds something that's potentially valuable, dopamine switches on, sending the message *Wake up. Pay attention. This is important.* It sends this message by creating the feeling of desire, and often excitement. The sensation of wanting is not a choice you make. It is a reaction to the things you encounter.

The man walking past the burger place smelled food, and although other priorities may have been floating around in his mind, dopamine gave him a near-overwhelming urge—he wanted that burger. Although the focus was different, this is the same mechanism that was working in our brains thousands of years ago. Imagine one of our ancestors walking along the savanna. It's a clear morning. The sun is coming up, the birds are singing, and everything is as it usually is. She walks along, looking without seeing, her mind wandering, when suddenly she stumbles upon a clump of bushes that are covered with berries. She's seen these bushes a dozen times before, but they never had berries on them. In the past her eyes slipped over these bushes, her thoughts somewhere else, but now she's paying attention. Her concentration sharpens as her eyes scan back and forth across the bushes, taking in all the details. Excitement wells up inside her. The future just became a little more secure because the bush with the dark green leaves makes fruit.

The desire circuit, powered by dopamine, has sprung into action.

She's going to remember this place where the berry bushes grow. From now on, whenever she sees this bush, a little dopamine will be released to make her more alert and to give her a hint of excitement, the better to motivate her to acquire this thing that can help her stay alive. An important memory has been formed: important because it's linked to survival, important because it was triggered by the release of dopamine. But what happens when dopamine gets out of control?

WHY WE LIVE IN A WORLD OF PHANTOMS

When Andrew saw an attractive woman, getting her into bed became the most interesting thing in his life. Everything else faded into dull gray. He typically met women in bars, and when he wasn't working, bars were where he wanted to be. Sometimes he told himself he was going to just relax and have a few beers. He liked the ambience, and there were times when he fought hard against the temptation to pick someone up. He knew that as soon as the sex was over, he would lose interest in the young woman, and he disliked that feeling. But in spite of knowing how things would turn out, he usually gave in.

After a while things got even worse. He was losing interest the moment the woman agreed to go home with him. The chase had come to an end, and everything was different. To his eyes, she even looked different, a transformation that occurred in the blink of an eye. By the time they got to his apartment he no longer wanted to have sex with her.

In a broad sense, saying something is "important" is another way of saying it's linked to dopamine. Why? Because among the many things it does, dopamine is an early-warning system for the appearance of anything that can help us survive. When something useful to our continued existence appears, we don't have to think about it. Dopamine makes us want it, right now. It doesn't matter if we're going to like it, or if we even need it at the moment. Dopamine doesn't care. Dopamine is like the little old lady who always buys toilet paper. It doesn't matter if she has a thousand rolls stacked in the pantry. Her attitude is *you can never have too much toilet paper.* So it is with dopamine, but instead of toilet paper, dopamine urges you to possess and accumulate anything that might help keep you alive.

This explains why the man on the diet wanted that hamburger even though he wasn't hungry. It explains why Andrew couldn't stop pursuing women even though he knew that in just a few hours, maybe in just a few minutes, it would make him unhappy. But it also explains more nuanced things; for instance, why we remember some names but not others. There are all sorts of tricks one can use to make remembering easier, such as using the person's name in conversation a few times.

But even if the name seems committed to memory, it almost always fades quickly. Important names—those of people who can affect our lives—are easier. The name of the person who flirted with you at the party will stay in your memory longer than the name of the person who ignored you. So will the name of the man who told you to set up an appointment to see him because he wants to give you a job—and his name will stick with you even more reliably if you're unemployed. Similarly, male rats remember the correct route through a maze more easily if there is a sexually receptive female at the other end. Sometimes the intensity of focus can be so great that your attention will get stuck on things that don't matter at the expense of things that do. A man who had a Beretta 9mm handgun pointed at his face during a robbery was asked to describe his assailant. He said, "I don't remember his face, but I can describe the gun."

Under more normal conditions, though, dopamine activation in the desire circuit triggers energy, enthusiasm, and hope. It feels good. In fact, some people spend the majority of their lives pursuing this feeling—a feeling of anticipation, a feeling that life is about to get better. You're about to eat a delicious dinner, see an old friend, make a big sale, or receive a prestigious award. Dopamine turns on the imagination, producing visions of a rosy future.

What happens when the future becomes the present—when the dinner is in your mouth or your lover is in your arms? The feelings of excitement, enthusiasm, and energy dissipate. Dopamine has shut down. Dopamine circuits don't process experience in the real world, only imaginary future possibilities. For many people it's a letdown. They're so attached to dopaminergic stimulation that they flee the present and take refuge in the comfortable world of their own imagination. "What will we do tomorrow?" they ask themselves as they chew their food, oblivious to the fact that they're not even noticing this meal they had so eagerly anticipated. *To travel hopefully is better than to arrive* is the motto of the dopamine enthusiast.

The future isn't real. It's made up of a bundle of possibilities that exist only in our minds. Those possibilities tend to be idealized—we usually don't imagine a mediocre outcome. We tend to think about the best of all possible worlds, and that makes the future more attractive. On the other hand, the present is real. It's concrete. It's experienced, not imagined, and that requires a different set of brain chemicals—the H&Ns, the here-and-now neurotransmitters. Dopamine makes us want things with a passion, but it's the H&Ns that allow us to appreciate them: the flavors, colors, textures, and aromas of a five-course meal, or the emotions we experience when we spend time with people we love.

WANTING VERSUS LIKING

The transition from excitement to enjoyment can be challenging. Think of buyer's remorse, the sense of regret that occurs after making a big purchase. Traditionally it has been attributed to the fear of having made the wrong choice, guilt over extravagance, or a suspicion of having been too influenced by the seller. In fact, it's an example of the desire circuit breaking its promise. It told you that if you bought that expensive car you'd be overcome with joy, and your life would never be the same. Except, once you became its owner, those feelings were neither as intense nor as long lasting as you had hoped. The desire circuit often breaks its promises—which is bound to happen, because it plays no role in generating feelings of satisfaction. It is in no position to make dreams come true. The desire circuit is, so to speak, just a salesman.

As we anticipate a desired purchase, our future-oriented dopamine system is activated and creates excitement. Once possession is achieved, the desired object moves from the *look up* extrapersonal space to the *look down* peripersonal space; from the future, distant realm of dopamine, and into the consummatory, near-body realm of H&N. Buyer's remorse is the failure of the H&N experience to compensate for the loss of dopaminergic arousal. If we made a wise purchase, it's possible that strong H&N gratification will make up for the loss of the dopamine thrill. Alternatively, another way to avoid buyer's remorse is to purchase something that triggers more dopaminergic expectation, for example, a tool, like a new computer that will boost productivity, or a new jacket that will make you look amazing the next time you go out.

Thus we see three possible solutions to buyer's remorse: (1) chase the dopamine high by buying more, (2) avoid the dopamine crash by buying less, or (3) strengthen the ability to transition from dopamine desire to H&N liking. In no case, though, is there any guarantee that the things we so desperately want will be things that we will enjoy having. Wanting and liking are produced by two different systems in the brain, so we often don't like the things we want. That's just what's going on in a scene from the sitcom *The Office* in which Will Ferrell, as temporary boss Deangelo Vickers, cuts into a large cake:

> **Deangelo:** I, for one, love the corners.
> *He slices off a corner, and eats it with his hand.*
> **Deangelo:** Why did I just do that? It's not even that good. I don't even want it. I had cake for lunch.
> *He throws what's left in his hand into the trash.*
> **Deangelo (sinking his fingers into the cake, and grabbing another handful):** No. You know what? I've been good. I deserve this.
> *He pauses, then:*
> **Deangelo:** What am I doing? C'mon, Deangelo!
> *He throws that handful away, too, then turns back to the cake. He leans down to the cake so he can yell at it.*
> **DeAngelo:** No! No!

Distinguishing between what we want and what we like can be difficult, but the disconnect is most dramatic when people become addicted to drugs.

HIJACKING THE DESIRE CIRCUIT

Since he spent so much time prowling for women, Andrew spent most of his free time in bars. When he was in college, he would go to keg parties where he drank until the early morning hours, so walking around with a beer in his hand felt natural. After graduation most of his drinking buddies moved on to other things. Alcohol no longer played a central role in their lives. But Andrew, for whom a bar was like home, kept at it. When he found someone he was interested in, he drank faster. Under the influence of a bright pair of eyes the world became a more exciting place, fueling his enjoyment of alcohol.

He knew that his drinking had become a problem when his morning hangovers made it hard for him to give his best at work. His sales began to slip, and his therapist advised him to take a break from drinking. The therapist recommended that Andrew try it for thirty days so he could experience what it was like to be sober. The therapist knew that if a heavy drinker can do this, he usually feels better—clearheaded, full of energy, better able to enjoy the simple pleasures in life—and that this feeling increases motivation for long-term sobriety. On the other hand, if a drinker can't achieve thirty days of sobriety, it's an indication that he no longer has full control of his drinking. That can be an eye-opening experience that may persuade a drinker to get alcohol out of his life.

Andrew tried it, and had no difficulty abstaining—except when he was in a bar looking for someone to sleep with. There was something about the place, something about the familiar experience of the chase, that triggered powerful cravings. His therapist became more concerned and felt Andrew met the criteria for an alcohol use disorder. He asked Andrew to try going to a few Alcoholics Anonymous meetings.

Andrew disagreed with the diagnosis. He was focused on overcoming his compulsion for anonymous sex. He was confident that if he could get that under control, he wouldn't have to go to bars anymore, and the alcohol problem would take care of itself. The therapy took a long time, and despite repeated discussions with his therapist, his drinking increased. Eventually, though, he attained his goal. He met someone who captured his interest, and to his delight the interest didn't fade. After a few fits and starts he completely gave up one-night stands. He no longer went out to bars much, but he was surprised to find that the drinking continued. The drinking had wormed its way into his brain, rewired his circuits, and now he couldn't stop.

Like a guided missile, addictive drugs hit the desire circuit with an intense chemical blast. No natural behavior can match that. Not food, not sex, not anything.

Alan Leshner, the former director of the National Institute on Drug Abuse, said that drugs "hijack" the desire circuit. They stimulate it far more intensely than natural rewards like food or sex, which affect the same brain-motivation system. That's why food and sex addictions have so much in common with addiction to drugs. Brain circuits that evolved for the crucial purpose of keeping us alive are taken over by an addictive chemical, and repurposed to enslave the addict that gets caught in its net.

Drug abuse is like cancer: it starts small but can quickly take over every aspect of a user's life. An alcoholic may start out as a moderate drinker. As he moves step by step from, say, a few beers on the weekend to a liter of vodka every day, other aspects of his life get swallowed up. At first he stops going to his son's baseball games so he can stay home and drink. After a while the parent–teacher conferences go, then all family activities, and last of all work, since that supplies money to buy the alcohol. But in the end even work goes. Like a tumor, the addiction has spread, and the alcoholic's entire life becomes focused exclusively on drinking. Was he making rational choices? From the outside it doesn't look like it.

But from the inside, where we see dopamine in action, it makes perfect sense.

The dopamine system evolved to motivate us to survive and reproduce. For most people there is nothing more important than staying alive and keeping their children safe. These are the activities that produce the largest dopamine surges. In a very literal way, large dopamine surges signal the need to react to life-and-death situations. *Take shelter. Find food. Protect your children.* These are tasks that hit the dopamine system hard. What could be more important?

To an addict, drugs are more important. At least that's the way it feels. That guided-missile dopamine blast overwhelms everything else. If making decisions is like weighing options on a balance, an addictive drug is an elephant sitting down on one side of the scale. Nothing else can compete.

An addict chooses drugs over work, family, everything. You think he's making irrational choices but his brain is telling him that his choices are perfectly logical. If someone offered you a choice between a meal at a nice restaurant, even the nicest restaurant in town, and a check for million dollars, it's ridiculous to think you'd choose dinner. That's exactly how an addict feels when choosing between, say, paying the rent and buying crack. He chooses the one that will lead to the bigger dopamine hit. The euphoria of crack cocaine is bigger than just about any experience you can name. That's rational from the point of view of desire dopamine, which is what drives the behavior of addicts.

Drugs are fundamentally different from natural dopamine triggers. When we're starving, there's nothing more motivating than getting food. But after we eat, the motivation for getting food declines because satiety circuits become active and shut down the desire circuit. There are checks and balances in place to keep everything stable. But there's no satiety circuit for crack. Addicts take drugs until they pass out, get sick, or run out of money. If you ask an addict how much crack he wants, there is only one answer: *more*.

Let's look at it from another angle. The goal of the dopamine system is to predict the future and, when an unexpected reward occurs, to send a signal that says, "Pay attention. It's time to learn something new about the world." In this way, circuits bathed in dopamine become malleable. They morph into new patterns. New memories are laid down, new connections are established. "Remember what happened," says the dopamine circuit. "This may be useful in the future."

What's the end result? You don't get surprised the next time the reward occurs. When you discovered the website that streamed your favorite music, it was exciting. But the next time you visited the site it wasn't. There's no longer any reward prediction error. Dopamine is not meant to be an enduring reservoir of joy. By shaping the brain to make surprising events predictable, dopamine maximizes resources, as it is supposed to do, but in the process, by eliminating surprise and extinguishing reward-prediction error, it suppresses its own activity.

But addictive drugs are so powerful that they bypass the complicated circuitry of surprise and prediction and artificially ignite the dopamine system. In this way, they scramble everything up. All that's left is a gnawing craving for more.

Drugs destroy the delicate balance that the brain needs to function normally. Drugs stimulate dopamine release no matter what kind of situation the user is in. That confuses the brain, and it begins to connect drug use to everything. After a while, the brain becomes convinced that drugs are the answer to all aspects of life. Feel like celebrating? Use drugs. Feeling sad? Use drugs. Hanging out with a friend? Use drugs. Feeling stressed, bored, relaxed, tense, angry, powerful, resentful, tired, energetic? Use drugs. People in twelve-step programs such as Alcoholics Anonymous say that addicts need to watch out for three things that might trigger craving and topple them into relapse: people, places, and things.

THE ADDICT WHO COULD NO LONGER MAKE HIS CLOTHES WHITE AND BRIGHT

Cues among addicts can be strange things. One former drug user had to avoid watching cartoons because his dealer printed cartoon characters on the drug packages he sold. Sometimes addicts don't even know what's triggering their craving. A struggling heroin addict found that he was overcome with craving every time he went to the grocery store. He had no idea why. It was causing havoc with his treatment. One day he and his counselor went on a field trip to the grocery store to try to figure out what was going on. The counselor told her patient to let her know as soon as the craving hit. They walked up and down each aisle, one by one, until suddenly the patient stopped and said, "Now." They were in the laundry detergent aisle, standing in front of a shelf full of bleach. Before he entered treatment, the addict had reused hypodermic needles by soaking them in bleach to avoid HIV infection.

THE REASON ADDICTS THINK SMOKING CRACK IS BETTER THAN SNORTING COCAINE

The ability to trigger dopamine in the desire circuit is what makes a drug addictive. Alcohol does it, heroin does it, cocaine does it, even marijuana does it. Not all drugs trigger dopamine to the same degree, though. The ones that hit dopamine the hardest are more addictive than ones that are more restrained. By triggering the release of more dopamine, the "hard hitters" also make the user feel more euphoric, and stimulate the most intense craving when the drug is gone. Intensity varies by drug. Pot smokers are generally less desperate to get more of the drug than cocaine addicts. But beneath all the differences is the commonality of the dopaminergic rush and subsequent craving.

Many factors account for the differences. The chemical structure of the molecules that make up each drug plays a large role; some chemicals are better than others at pushing dopamine along its path. But there are other considerations as well. For example, the crack cocaine that users smoke is essentially the same molecule as powder cocaine that users snort, but crack is far more addictive—so much so that when crack became widely available in the 1980s, it took the world of recreational drug use by storm.

What's so "great" about crack that allowed it to take over the cocaine market, and chemically enslave thousands of people? From a scientific perspective, the answer is simple: the rate of *onset of action.*

Consider a drug such as alcohol that triggers the release of dopamine. The faster it gets into the brain, the higher it will make its user. In Figure 2 the horizontal axis shows how much time has gone by and the vertical axis shows how much drug has gotten into a user's brain. If someone is sipping a glass of Chardonnay, the graph will gently rise to the right. On the other hand, if that same person were to start taking shots of vodka, the graph would show a steep slope that quickly shoots upward.

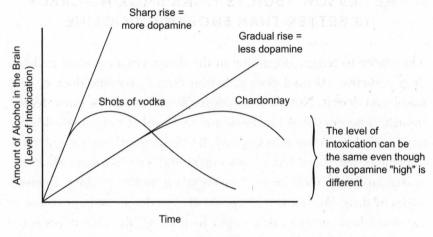

Figure 2

The slope of the line indicates how quickly the level of the drug—in this case, alcohol—is rising in the brain. And the faster the rise, the more dopamine release, the more euphoria, and the more craving down the road.

That's why smoking crack is more appealing than snorting powder cocaine: smoking produces a faster, larger dopamine rush. Regular cocaine can't be smoked; the heat destroys it. Transforming it into crack makes it smokable, so the drug gets in the body through the lungs instead of the nose. That makes a big difference.

When powdered cocaine flies up into the nose, it lands on the nasal mucosa, the red lining inside your nose. It's red because the blood vessels are at the surface. Cocaine enters the bloodstream through these vessels, but it's not very efficient; there isn't much room available in there. Sometimes when a user snorts a line of cocaine, some of the powder never makes it into their system because there's not enough space for it on the surface of the mucosa.

That's not to say that snorting cocaine isn't dangerous and addictive, but there's a way to make it even more dangerous and more addictive: smoking it. Smoking cocaine as crack makes the process more efficient. Unlike the nasal mucosa, the surface area of the lungs is huge. Filled with hundreds of millions of tiny air sacs, the surface area is equal to one side of a tennis court. There's plenty of room there, and when the vaporized cocaine hits the lungs, it goes right into the bloodstream and up to the brain. It's a steep slope—a sudden burst—and a big hit to the dopamine system.

The link between a rapidly rising blood level and dopamine release is also why addicts progress to injecting drugs into their veins. Other routes of administration no longer give them the thrill they're after. Injecting drugs is scary, though, and is a clear sign of an addict, so the stigma and fear of the needle may stop many of them from progressing further. Unfortunately, smoking gets the drug into the brain about as fast as intravenous injection. Smoking also lacks the stigma associated with needles. As a result, many would-be casual users of cocaine progress to life-destroying addiction. The same thing happened with methamphetamine when it became available in smokable form.

DRUNK VERSUS HIGH: WHAT'S THE DIFFERENCE?

There's a big difference between being high and being drunk, but not everyone knows that. Even fewer understand why.

An evening of drinking feels best at the start. The level of alcohol is rising rapidly, and that feels good—it's *dopaminergic euphoria*, directly related to how fast the alcohol gets into the brain. As the night goes on, though, the rate of increase slows down, and dopamine turns off. Euphoria gives way to drunkenness. The early stage of rising levels of alcohol might be characterized by increased energy, excitement, and pleasure. Intoxication, on the other hand, is characterized by sedation, poor coordination, slurred speech, and bad judgment. The speed with which alcohol gets into the brain determines how high a drinker feels. It's the total amount of alcohol consumed, regardless of whether it's fast or slow, that determines the level of intoxication.

Inexperienced drinkers get the two confused. They start drinking, push their blood alcohol level up, and experience the pleasures of dopamine release, then mistakenly believe that the pleasure is the pleasure of intoxication. So they keep drinking more and more, trying in vain to get the rush back. It ends badly, often bent over a toilet.

Some people figure this out on their own. A woman seen at a cocktail party said she always had more fun with mixed drinks than with beer. At first this appeared to be nonsense, because alcohol is alcohol whether it comes from a beer or a daiquiri. But science validates the woman's experience. A mixed drink is more concentrated, and it's usually sweetened with sugar, so people tend to drink it faster. Mixed drinks usually contain more alcohol than beer or wine. Therefore a mixed drink delivers a lot of alcohol fast, a burst of dopaminergic stimulation, as opposed to an evening of slowly increasing intoxication. This woman wanted elation, not inebriation, so of course the mixed drinks let her have a better time. She was getting a dopamine hit from a few cocktails that an evening of many beers couldn't deliver.

THE CRAVING THAT NEVER STOPS

Although craving never stops as long as an addict keeps using drugs, the brain gradually loses its ability to deliver the high—the desire circuit simply reacts less and less, so much so that they might as well replace the drug with salt water.[1]

---

1   When scientists injected long-time cocaine users with a stimulant similar to cocaine, they released 80 percent less dopamine than healthy people given the same drug. The dopamine released by the addicts was about the same amount that the scientists saw when they injected a placebo—an inactive substance, such as salt water.

Patrick Kennedy, the former U.S. representative for Rhode Island's 1st Congressional District, and son of the late Massachusetts senator Ted Kennedy, understands the diminishing stimulation of drug use. Arguably the foremost advocate for brain research and improved mental health services in the United States, he himself struggled with addictions and mental illness, publicly acknowledging his problems after he drove into a barricade at the U.S. Capitol in the middle of the night. In a *60 Minutes* interview with Lesley Stahl he spoke of the need to use, even in the absence of pleasure.

> *There's no partying there. There's no enjoyment. This is about relieving the pain. People have this mistaken notion that you get high. What you're really getting is relief from the low.*

This is why, even if an addict uses so much cocaine (or heroin or alcohol or marijuana) that it no longer leads to feeling high, he will continue to use it.

Remember the happy surprise of the bakery with the delicious croissants and coffee? You were walking along expecting nothing, something good appeared, and your dopamine system leaped into action—hence your "prediction" was wrong, and you experienced the burst of dopamine from reward prediction error. You started going to that bakery every day. Now imagine that you're waiting in line for your morning coffee and croissant, and all of a sudden your phone rings. It's your boss. There's a crisis at work. Drop whatever you're doing, she says, and get to the office right away. Assuming you're a conscientious person, you'll leave the bakery empty-handed, feeling resentful and deprived.

Now let's say it's Saturday night, and an addict's brain is expecting the usual Saturday-night "treat," cocaine, but it doesn't come. Just like the croissant-deprived office worker, the drug-deprived addict will feel resentful and deprived.

When an expected reward fails to materialize, the dopamine system shuts down. In scientific terms, when the dopamine system is at rest, it fires at a leisurely three to five times per second. When it's excited, it zooms up to twenty to thirty times per second. When an expected reward fails to materialize, the dopamine firing rate drops to zero, and that feels terrible.

That's why a dopamine shutdown makes you feel resentful and deprived. It's how a recovering drug addict feels every day as he struggles to get clean and sober. It takes an enormous amount of strength, determination, and support to overcome addiction. Don't mess with dopamine. It hits back hard.

DESIRE IS PERSISTENT, BUT HAPPINESS IS FLEETING

Giving in to craving doesn't necessarily lead to pleasure because wanting is different from liking. Dopamine makes promises that it is in no position to keep. "If you buy these shoes, your life will change," says the desire circuit, and it just might happen, but not because dopamine made you feel it.

Dr. Kent Berridge, a professor of psychology and neuroscience at the University of Michigan, is a pioneering figure in the process of disentangling dopamine desire circuits from here-and-now liking circuits. He found that when a rat tastes a sugar solution, it signals liking by licking its lips. In contrast, it expresses wanting by consuming more of the sweet liquid. When he injected a chemical into a rat's brain that boosted dopamine, it consumed more sugar water, but didn't show any increased signs of liking. On the other hand, when he injected an H&N booster, he was able to triple the lip-smacking liking response. All of a sudden the sugar water became far more delicious.

In an interview with *The Economist*, Dr. Berridge noted that the dopamine desire system is powerful and highly influential in the brain, whereas the liking circuit is tiny, fragile, and much harder to trigger. The difference between the two is the reason that "life's intense pleasures are less frequent and less sustained than intense desire."

Liking involves different circuits in the brain, and uses the H&N chemicals, not dopamine, to send messages. In particular, liking relies on the same chemicals that promote the long-term satisfaction of companionate love: endorphins and endocannabinoids. Because opioid drugs such as heroin and OxyContin scramble both the desire circuit and the liking circuit (where dopamine acts and where endorphin acts), they are among the most addictive drugs there are. Marijuana is similar. It also interacts with both circuits, stimulating dopamine as well as the endocannabinoid system. This dual effect leads to unusual results.

Boosting dopamine can lead to enthusiastic engagement with things that would otherwise be perceived as unimportant. For example, marijuana users have been known to stand in front of a sink, watching water drip from the faucet, captivated by the otherwise mundane sight of the drops falling over and over again. The dopamine-boosting effect is also evident when marijuana smokers get lost in their own thoughts, floating aimlessly through imaginary worlds of their own creation. On the other hand, in some situations marijuana suppresses dopamine, mimicking what H&N molecules tend to do. In that case, activities that would typically be associated with wanting and motivation, such as going to work, studying, or taking a shower, seem less important.

IMPULSIVENESS AND THE DOWNWARD SPIRAL

Many of the decisions that addicts make, particularly the harmful decisions, are impulsive. Impulsive behavior occurs when too much value is placed on immediate pleasure and not enough on long-term consequences. Desire dopamine overpowers the more rational parts of the brain. We make choices that we know are not in our best interest, but we feel powerless to resist. It's as if our free will has been compromised by an overwhelming urge for immediate pleasure; perhaps it's a bag of potato chips when we're on a diet, or splurging on an expensive night out that we can't really afford.

Drugs that boost dopamine can also boost impulsive behavior. A cocaine addict once said, "When I do a line of cocaine, I feel like a new man. And the first thing that new man wants is another line of cocaine." When the addict stimulates his dopamine system, his dopamine system responds by demanding more stimulation. That's why most cocaine addicts smoke cigarettes when they use cocaine. Like cocaine, nicotine stimulates additional dopamine release, but it's cheaper and easier to get.

Nicotine, in fact, is an unusual drug because it does very little except trigger compulsive use. According to researcher Roland R. Griffiths, PhD, professor of psychiatry and behavioral sciences at the Johns Hopkins University School of Medicine, "When you give people nicotine for the first time, most people don't like it. It's different from many other addictive drugs, for which most people say they enjoy the first experience and would try it again." Nicotine doesn't make you high like marijuana or intoxicated like alcohol or wired up like speed. Some people say it makes them feel more relaxed or more alert, but really, the main thing it does is relieve cravings for itself. It's the perfect circle. The only point of smoking cigarettes is to get addicted so one can experience the pleasure of relieving the unpleasant feeling of craving, like a man who carries around a rock all day because it feels so good when he puts it down.

Addiction arises from the chemical cultivation of desire. The delicate system that tells us what we like or dislike is no match for the raw power of dopaminergic compulsion. The feeling of wanting becomes overwhelming and utterly detached from whether the object of desire is anything we really care for, is good for us, or might kill us. Addiction is not a sign of weak character or a lack of willpower. It occurs when the desire circuits get thrown into a pathological state by overstimulation.

Prod dopamine too hard and too long, and its power comes roaring out. Once it has taken charge of a life, it is difficult to tame.

THE PARKINSON'S PATIENT WHO LOST HIS HOME TO VIDEO POKER

Recreational drugs aren't the only ones that stimulate dopamine. There are prescription drugs that do it as well, and when they hit the desire circuit too hard, strange things can happen. Parkinson's disease is an illness of dopamine deficiency in a pathway that's responsible for controlling muscle movements. Or, to put it more simply, it's how we translate our inner world of ideas into action, the way we impose our will upon the world. When there is not enough dopamine in this circuit, people become stiff and shaky, and they move slowly. The treatment is to prescribe drugs that boost dopamine.

Most people who take these drugs do just fine, but about one in six patients gets into trouble with high-risk, pleasure-seeking behavior. Pathological gambling, hypersexuality, and compulsive shopping are the most common ways the excessive dopamine stimulation is seen. To explore this risk, British researchers gave a drug called L-dopa to fifteen healthy volunteers. L-dopa is made into dopamine inside the brain, and can be used to treat Parkinson's disease. They gave another fifteen volunteers a placebo. Nobody knew who got the drug and who got the fake pill.

After they took the pills, the volunteers were given the opportunity to gamble. The researchers found that the participants who took the dopamine-boosting pill placed larger and riskier bets than those who took the placebo. The effect was more pronounced in men than in women. The researchers periodically asked the participants to rate how happy they were. There was no difference between the two groups. The enhanced dopamine circuit boosted impulsive behavior, but not satisfaction—it boosted the wanting, but not the liking.

When the scientists used powerful magnetic fields to look inside their participants' brains, they found yet another effect: the more active the dopamine cells were, the more money the volunteers expected to win.

It's not uncommon for people to deceive themselves in this way. There are few things we encounter in daily life that are more unlikely than winning the lottery. A person is more likely to have identical quadruplets, or be killed by a vending machine tipping over. It's over a hundred times more likely that a person will be struck by lightning than win the lottery. Yet millions of people buy tickets. "Someone has to win," they say. A more sophisticated dopamine enthusiast expressed his devotion to the lottery in this way: "It's hope for a dollar."

Expecting to win the lottery may be irrational, but far more severe distortions of judgment can occur when people take dopamine-boosting medicines every day:

On March 10, 2012, lawyers for Ian,[2] a sixty-six-year-old resident of Melbourne, Australia, filed a statement of claim in federal court. He was suing the drug manufacturer Pfizer, claiming that their Parkinson's medication, Cabaser, made him lose everything he had.

He was diagnosed with Parkinson's disease in 2003. His doctor prescribed Cabaser, and in 2004 Ian's dose was doubled. That's when the problems began. He started gambling heavily on video poker machines. He was retired, and received a modest pension of about $850 per month. Each month he fed the entire sum into the machines, but it wasn't enough. To pay for his compulsion, he sold his car for $829, pawned much of what he owned for $6,135, and borrowed $3,500 from friends and family. Next, he took out loans for over $50,000 from four financial institutions, and on July 7, 2006, he sold his home.

In all, this man of modest means gambled away over $100,000. He was finally able to stop in 2010, when he read an article about the link between Parkinson's medication and gambling. He stopped taking Cabaser, and the problem went away.

Why do some people who take Parkinson's medication engage in destructive behavior, but most do not? It's possible they were born with a genetic vulnerability. People who gambled frequently in the past are more likely than others to experience out-of-control gambling after they start Parkinson's medication, suggesting there are certain personality features that put people at risk.

Another risk of Parkinson's medication is hypersexuality. A Mayo Clinic case series—the tracking of patients with a certain type of illness or treatment—described a fifty-seven-year-old man treated with L-dopa who "would have sexual intercourse twice daily and, when possible, even

---

2 To protect privacy, we have disguised or created composites of individuals and their cases throughout the book.

more often. Both he and his wife worked full time, and because of her busy schedule, she found it difficult to satisfy him." After he retired at age sixty-two, things became worse. He made sexual advances to two young ladies in his extended family as well as to women in the neighborhood. Eventually, his wife had to leave her job to attend to his sexual urges.[3]

Yet another patient expressed his hypersexuality by spending hours every day in internet adult chat rooms—but even otherwise healthy people taking no medication at all are susceptible to the dopamine call of pornography, supercharged by the internet.

Of course, you don't need Parkinson's medication coursing through your brain to have your life upended by sexual obsession. Consider the fearsome triad of dopamine, technology, and porn.

MORE, MORE, MORE: DOPAMINE AND THE POWER OF PORNOGRAPHY

Noah was a twenty-eight-year-old man who sought help because he was unable to stop viewing pornography. He grew up in a Catholic household, and the first time he was exposed to pornography was at the age of fifteen. He was on the internet searching for something unrelated when he ran across a picture of a naked woman. He said from that moment he was hooked.

At first, things weren't too bad. He was accessing the internet over a dial-up modem and "it took ages for the pictures to load." He was lucky. Technology was limiting his daily dose. He described the pictures he started with as "tame." Over time, both of these would change. Broadband allowed him to access pictures instantly, and he could now add video to his daily routine. Tame material gave way to depictions of more extreme acts as his tolerance for pornographic thrills increased.

He considered his behavior to be a sin, a moral failing, and he used his relationship with the church to get his compulsion under control. He went

---

3   This problem primarily affects men, but women are not immune. In the Mayo Clinic series of thirteen patients, two were female, both single and sexually abstinent prior to starting treatment.

to confession on a regular basis, and received emotional support that helped him cut back his viewing habits. But when his work assigned him to an overseas branch, everything fell apart. Unable to speak the local language, he became socially isolated, and his compulsion flared up worse than ever. He said, "What makes this so hard is the inner struggle, the conflict within. It's a war against yourself." Feeling completely out of control, he no longer believed it was a strictly moral failing. "I need to fight this on a chemical level because at some point I want to get married."

Thanks to the internet, sexually graphic material is more easily available than ever before. Some people maintain that it is possible to become addicted to pornography, even for otherwise healthy people taking no medication at all. In 2015 the *Daily Mail* claimed that as many as one in twenty-five young adults in the United Kingdom were believed to be sex addicts.

A reporter from the newspaper spoke to researchers at the University of Cambridge who described experiments in which they had placed young men in brain scanners, and then piped in pornographic videos for them to watch. As expected, their dopamine circuits lit up. The circuits went back to normal when ordinary videos were displayed.

The scientists put other volunteers in front of a computer, and found that of all the content on the internet, pictures of undressed women were most likely to make young men click compulsively. They also discovered that showing people "highly arousing sexual pictures" was distracting when they were trying to pay attention to something else. (Amateur scientists can try this experiment at home.) At the end of the study they concluded that compulsive sexual behavior was fueled by easy access to sexual images on the internet.

THE POWER OF EASY ACCESS

When it comes to addiction, easy access matters. More people get addicted to cigarettes and alcohol than to heroin, even though heroin hits the brain in a way that is more likely to trigger addiction. Cigarettes and alcohol are a larger public health problem because they are so easy to obtain. In fact, the most effective way to reduce the problems caused by these substances is to make it more difficult to get them.

We've all seen "quit smoking" advertisements on buses and subways. They don't work. We've heard about school programs that teach kids to say no to drugs and alcohol. In many cases drug and alcohol use go up after these programs because they pique the curiosity of the adolescent students. The only thing that has been shown to work consistently is raising taxes on these products and placing limits on where and when they can be sold. When these measures are taken, use goes down.[4]

As barriers to the use of tobacco have gone up, barriers to pornography have gone down. In the past, getting sexually explicit pictures was something of an ordeal. People had to muster the courage to walk into a drugstore, pick up a magazine, and then hope the cashier wasn't a member of the opposite sex. Today, pornographic pictures and videos can be had in seconds and in complete privacy. There are no barriers of embarrassment or shame.

We don't yet know if compulsive viewing of pornography is exactly the same as drug addiction, but they have things in common. As with drug addiction, people who become trapped in a cycle of excessive pornography use spend more and more time pursuing this activity—sometimes many hours every day. They abandon other activities so they can focus on adult internet sites. Sexual relations with their partners tend to become less frequent and less satisfying. One young man gave up dating completely. He said that he'd rather look at pornography than go out with a real woman because the women in the pictures never demanded anything of him, and never said no.

---

4   Raising the price of cigarettes and alcohol is controversial, though, especially with regard to cigarettes. Fewer and fewer people smoke. Those who persist tend to be poor and less educated. As a result, increases in cigarette taxes hit them the hardest. This is the opposite of a tax system that shifts more of the burden onto those who are better able to afford it. Advocates who defend this strategy argue that the pain caused by raising taxes on the poor is counterbalanced by reducing their risk of getting cancer, emphysema, and heart disease.

As with drugs, habituation can also occur with pornography, in which the starting "dose" no longer works as well. When sex addicts were repeatedly shown the same sexual images, their interest diminished. The activity measured in their dopamine circuits also decreased as the images were shown over and over. The same thing happened to healthy males who were repeatedly shown the same pornographic video. When they were shown a new video, their dopamine systems revved up again. This experience of a dopamine rush, followed by a dopamine drop (repeated images), followed by another dopamine rush (new images), pushes addicts to continually seek out fresh material, which may explain why browsing internet sex sites can become compulsive. It's hard to resist the demands of dopamine circuits, especially with something as evolutionarily important as sex. The researchers who performed the study also identified a wanting/liking divide similar to what is seen in drug addiction: "Sex addicts showed higher levels of desire when watching pornography, but did not necessarily rate the explicit videos higher in their 'liking' scores."

ARE VIDEO GAMES ADDICTIVE, TOO?

It's not just pornography that can ensnare computer users. Some scientists claim that video games can also be addicting. In certain ways, video games are similar to casino games. Like slot machines, video games surprise players with unpredictable rewards. They do more than that, though, which can make them even more potent agents of dopamine release. In researching this problem, psychologist Douglas Gentile of Iowa State University found that nearly one in ten gamers ages eight to eighteen are addicted, causing family, social, school, or psychological damage because of their video game playing habits—a rate of addiction more than five times higher than that among gamblers, according to the National Research Council on Pathological Gambling. What accounts for this large difference in how many users get addicted?

Part of the difference is that the video gamers Gentile studied were adolescents. It's unusual for adults to experience serious negative consequences from playing video games. Adolescent brains, however, have not yet fully developed, so adolescents may act like adults with brain damage. The biggest difference in the adolescent brain is in the frontal lobes, which don't completely develop until their early twenties. That's a problem because it's the frontal lobes that give adults good judgment. They act like a brake, warning us when we're about to do something that might not be such a good idea. Without fully functioning frontal lobes, adolescents act impulsively, and are at greater risk of making unwise decisions, even when they know better.

There's more to it than that, though. Video games are more complex than slot machines, so there are more opportunities for programmers to bake in features that trigger dopamine release in order to make it hard to stop playing.

Video games are all about imagination. They immerse us in a world where our fantasies can come true, where reality-shunning dopamine can bask in endless possibilities. We can explore environments that constantly change, ensuring that the surprises never end. We may start off in the desert, progress to a rain forest, then a dark alley in a gritty urban hell, then suddenly we're on a rocket, hurtling toward an alien world.

Players do more than just explore, though. Video games are about progress. They're about making the future better than the present. Gamers progress through levels while increasing their strength and abilities. It's a dopamine dream come true. To keep progress front and center in a gamer's mind, the screen constantly displays the accumulating points or growing progress bars so players never forget. They have to keep pursuing *more.*

Video games are full of rewards. Gamers collect coins, hunt for treasure, or maybe capture magic unicorns to progress to the next level. Players' expectations are constantly kept off balance because they never know where the next reward will be. Some games make you kill monsters to earn points; others make you look inside treasure chests.

When a player opens a newly discovered chest, it may contain what he's looking for, but not always. If you needed to collect, say, seven gems, and every chest you opened contained a gem, it would be completely predictable. There would be no surprises, no reward prediction errors, no dopamine. If, on the other hand, you had to open a thousand chests to find a single gem, it would be so frustrating that everyone would give up. How does a game developer decide what percentage of chests should contain a gem? The answer is data. Lots of data.

Online games are constantly collecting information about players. How long do they play? When do they quit? What kinds of experiences make them play longer? What kinds make them give up? According to gaming theorist Tom Chatfield, the biggest online games have accumulated billions of data points about their players. They know exactly what lights up dopamine, and what turns it off—though game designers are not thinking of these events as dopamine triggers, but simply as "what works."

So, what do the data tell us about the ideal portion of treasure chests that should contain gems? It turns out that 25 percent is the magic number. That's what keeps people playing the longest. And there's no reason why the other 75 percent should be empty. Game developers put low-value rewards in the non-gem chests so every single one will contain a surprise. Maybe it's a small coin. Maybe it's a new scope for your rifle. Maybe it's a pair of sunglasses that will make your online character look cool. Or maybe it's something so powerful that it opens up completely new ways to interact with the game. Chatfield tells us that a reward like that should be found in only one out of a thousand treasure chests. (By the way, the game probably won't let you progress to the next level with only those seven gems. The billions of data points tell us that fifteen is the optimal number for getting people to play as long as possible.)

It's worth mentioning that there are also H&N pleasures in video games that contribute to their appeal. Many games let you play with friends. The pleasure we get when we socialize for no other reason than the enjoyment of the company of others is an H&N experience. On the other hand, when we get together to accomplish a shared goal, it's dopaminergic because we're working toward a better future (even if it's just capturing the enemy's base). Video games provide both types of social pleasure.

Many video games are also beautiful, another way of stimulating H&N delight. Some of them are, in fact, astonishing because enormous resources have been poured into amassing talented people to create them. The *Los Angeles Times* reported that developing the online game *Star Wars: The Old Republic* required more than eight hundred people on four continents at a cost of over $200 million. The world of the game is vast. Working through all the story lines would require 1,600 hours of play. Spending that much money to create a game is risky, but there's the potential for a big payoff. *Grand Theft Auto*, one of the most successful video game series, had sales of $1 billion in just three days for its fifth-generation release. Americans spend more than $20 billion per year on video games; they spent only half that much on movie tickets in 2016, the biggest U.S. box office year in history.

DOPAMINE VERSUS DOPAMINE

It's natural to confuse wanting and liking. It seems obvious that we would want the things that we will like having. That's how it would work if we were rational creatures, and despite all evidence to the contrary, we persist in thinking that we *are* rational creatures. But we're not. Frequently we want things that we don't like. Our desires can lead us toward things that may destroy our lives, such as drugs, gambling, and other out-of-control behaviors.

The dopamine desire circuit is powerful. It focuses attention, motivates, and thrills. It has a profound influence over the choices we make. Yet it isn't all-powerful. Addicts get clean. Dieters lose weight. Sometimes we switch off the TV, get off the couch, and go for a run. What kind of circuit in the brain is powerful enough to oppose dopamine? Dopamine is. Dopamine opposing dopamine. The circuit that opposes the desire circuit might be called the *dopamine control circuit*.

You may recall that in many situations, future-focused dopamine opposes the activity of the H&N circuits and vice versa. If you're thinking about where to go for dinner, you're probably not appreciating the taste, smell, and texture of the sandwich you're eating for lunch. But there's also opposition within the future-oriented dopamine system itself.

Why would the brain develop circuits that work against each other? Wouldn't it make more sense to have everyone pulling together, so to speak? In fact, no. Systems that contain opposing forces are easier to control. That's why cars have both an accelerator and a brake, and why the brain uses circuits that counter each other.

Not surprisingly, the dopamine control circuit involves the frontal lobes, the part of the brain that is sometimes called the neocortex because it evolved most recently. It's what makes human beings unique. It gives us the imagination to project ourselves further into the future than the desire circuit can take us, so we can make long-term plans. It's also the part that allows us to maximize resources in that future by creating new tools and using abstract concepts; concepts that rise above the here-and-now experience of the senses, like language, mathematics, and science. It's intensely rational. It doesn't feel, because emotion is an H&N phenomenon. As we will see in the next chapter, it's cold, calculating, and ruthless, doing whatever it takes to reach its goal.

FURTHER READING

Pfaus, J. G., Kippin, T. E., & Coria-Avila, G. (2003). What can animal models tell us about human sexual response? *Annual Review of Sex Research*, 14(1), 1–63.

Fleming, A. (2015, May–June). The science of craving. *The Economist 1843*. Retrieved from https://www.1843magazine.com/content/features/wanting -versus-liking

Study with "never-smokers" sheds light on the earliest stages of nicotine dependence. (2015, September 9). *Johns Hopkins Medicine*. Retrieved from https:// www.hopkinsmedicine.org/news/media/releases/study_with_never_ smokers_sheds_light_on_the_earliest_stages_of_nicotine_dependence

Rutledge, R. B., Skandali, N., Dayan, P., & Dolan, R. J. (2015). Dopaminergic modulation of decision making and subjective well-being. *Journal of Neuroscience*, 35(27), 9811–9822.

Weintraub, D., Siderowf, A. D., Potenza, M. N., Goveas, J., Morales, K. H., Duda, J. E., . . . Stern, M. B. (2006). Association of dopamine agonist use with impulse control disorders in Parkinson disease. *Archives of Neurology*, 63(7), 969–973.

Moore, T. J., Glenmullen, J., & Mattison, D. R. (2014). Reports of pathological gambling, hypersexuality, and compulsive shopping associated with dopamine receptor agonist drugs. *JAMA Internal Medicine*, 174(12), 1930–1933.

Ian W. v. Pfizer Australia Pty Ltd. Victoria Registry, Federal Court of Australia, March 10, 2012.

Klos, K. J., Bower, J. H., Josephs, K. A., Matsumoto, J. Y., & Ahlskog, J. E. (2005). Pathological hypersexuality predominantly linked to adjuvant dopamine agonist therapy in Parkinson's disease and multiple system atrophy. *Parkinsonism and Related Disorders*, 11(6), 381–386.

Pickles, K. (2015, November 23). How online porn is fueling sex addiction: Easy access to sexual images blamed for the rise of people with compulsive sexual behaviour, study claims. *Daily Mail*. Retrieved from http://www. dailymail.co.uk/health/article-3330171/How-online-porn-fuelling-sex -addiction-Easy-access-sexual-images-blamed-rise-people-compulsive -sexual-behaviour-study-claims.html

Voon, V., Mole, T. B., Banca, P., Porter, L., Morris, L., Mitchell, S., . . . Irvine, M. (2014). Neural correlates of sexual cue reactivity in individuals with and without compulsive sexual behaviors. *PloS One*, 9(7), e102419.

Dixon, M., Ghezzi, P., Lyons, C., & Wilson, G. (Eds.). (2006). *Gambling: Behavior theory, research, and application*. Reno, NV: Context Press.

National Research Council. (1999). *Pathological gambling: A critical review*. Chicago: Author.

Gentile, D. (2009). Pathological video-game use among youth ages 8 to 18: A national study. *Psychological Science*, 20(5), 594–602.

Przybylski, A. K., Weinstein, N., & Murayama, K. (2016). Internet gaming disorder: Investigating the clinical relevance of a new phenomenon. *American Journal of Psychiatry*, 174(3), 230–236.

Chatfield, T. (2010, November). Transcript of "7 ways games reward the brain." Retrieved from https://www.ted.com/talks/tom_chatfield_7_ways_games _reward_the_brain/transcript?language=en

Fritz, B., & Pham, A. (2012, January 20). Star Wars: The Old Republic—the story behind a galactic gamble. Retrieved from http://herocomplex.latimes.com/ games/star-wars-the-old-republic-the-story-behind-a-galactic-gamble/

Nayak, M. (2013, September 20). Grand Theft Auto V sales zoom past $1 billion mark in 3 days. Reuters. Retrieved from http://www.reuters.com/article/ entertainment-us-taketwo-gta-idUSBRE98J0O820130920

Ewalt, David M. (2013, December 19). Americans will spend $20.5 billion on video games in 2013. *Forbes*. Retrieved from https://www.forbes.com/ sites/davidewalt/2013/12/19/americans-will-spend-20-5-billion-on-video -games-in-2013/#2b5fa4522c1e

Chapter 3

DOMINATION

How far will you go?

In which dopamine drives us to overcome complexity, adversity, emotion, and pain so we can control our environment.

PLANNING AND CALCULATION

Just wanting rarely gets you much of anything. You have to figure out how to obtain it, and whether it is worth having at all. In fact, when we do things without thinking about *how* and *what next*, failure is not even the worst possible outcome. The results can range from a little overeating all the way to reckless gambling, drug abuse, and worse.

Desire dopamine makes us want things. It is the source of raw desire: *give me more.* But we're not at the ungoverned mercy of our desire. We also have a complementary dopamine circuit that calculates what sort of *more* is worth having. It gives us the ability to construct plans—to strategize and dominate the world around us to get the things we want. How does a single chemical do both things? Think of rocket fuel that powers the main engines of a spaceship. The same fuel that pushes the rocket forward can be redirected to drive directional thrusters to steer the ship, as well as retrorockets to slow it down. It all depends on the path the fuel takes before it's ignited—different functions, but all working together to get the spaceship to its destination. In a similar way, dopamine moving through different brain circuits yields different functions, too, and toward a common end: a relentless focus on enhancing the future.

Urges come from dopamine passing through the mesolimbic circuit, which we call the dopamine desire circuit. Calculation and planning—the means of dominating situations—come from the mesocortical circuit, which we will call the dopamine control circuit (Figure 3). Why call it the control circuit? Because its purpose is to manage the uncontrolled urges of desire dopamine, to take that raw energy and guide it toward profitable ends. Also, by using abstract concepts and

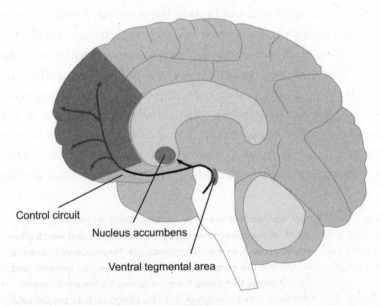

Control circuit

Nucleus accumbens

Ventral tegmental area

Figure 3

forward-looking strategies, it allows us to gain control over the world around us, and dominate our environment.[1]

In addition, the dopamine control circuit is the source of imagination. It lets us peer into the future to see the consequences of decisions we might make right now, and thus allows us to choose which future we prefer. Finally, it gives us the ability to plan how to make that imaginary future a reality. Like the desire circuit, which only cares about things we do not have, control dopamine works in the unreal world of the possible. The two circuits begin in the same place, but the desire circuit ends in a part of the brain that triggers excitement and enthusiasm, while the control circuit goes to the frontal lobes, a part of the brain that specializes in logical thinking.

In this way, both circuits give us the capacity to consider "phantoms"—things that don't physically exist. For desire dopamine, those phantoms are things we wish to have but don't have right now—things we want in the future. For control dopamine, the phantoms are the building blocks of imagination and creative thought: ideas, plans, theories, abstract concepts such as mathematics and beauty, and worlds yet to be.

Control dopamine carries us beyond the primitive *I want* of desire dopamine. It gives us tools to comprehend, analyze, and model the world around us, so we can extrapolate possibilities, compare and contrast them, then craft ways to achieve our goals. It is an extended and complex execution of the evolutionary imperative: to secure as many resources as possible. In contrast, desire dopamine is the kid in the back seat shouting for his parents to "Look! Look!" every time he sees a McDonald's, a toy store, or a puppy on the sidewalk. Control dopamine is the parent at the wheel, hearing each request and considering

---

1   We'll be using the term *environment* in a different way than it is commonly used. When most people think of the environment, they think of the natural world, often as something we need to protect, as in *environmentalism*. Neuroscientists use the word to refer to everything in the external world that influences our behavior and health, as opposed to influences that come from our genes. So the environment includes not only mountains, trees, and grass, but also things such as people, relationships, food, and shelter.

whether it's worth stopping for—and deciding what to do if he pulls over. Control dopamine takes the excitement and motivation provided by desire dopamine, evaluates options, selects tools, and plots a strategy to get what it wants.

For example, a young man is planning to buy his first car. If all he had was desire dopamine, he would buy the first one that caught his eye. But since he also has control dopamine, he's able to refine that impulse. There are any number of reasons to prefer one car over another; let's say this young man is thrifty, and wants the best car he can afford at the lowest price. Tapping into desire dopamine energy, he spends hours on the internet, poring over car review sites and developing negotiation strategies. He wants to know every detail he can so he can maximize the value of his purchase. When he sits down with the car dealer, he is so well prepared that nothing will take him by surprise. He feels good: he has dominated the car-buying situation by mastering all available information.

Consider a woman on her way to work. She drives to the train station, taking a roundabout route that avoids the morning rush hour traffic. When she gets to the station, she navigates to an unoccupied corner of the parking garage that few people know about, and easily finds a place to park. She waits on the platform at the precise spot where she knows the doors to the commuter train will open, putting her at the front of the line, ready to get one of the remaining empty seats for the long ride to the city. She feels good: she has dominated her commute.

It's fun figuring out things, and it's fun carrying out the strategies developed to "game" the intricacies of car buying and the daily trip to work. Why? As always, the function of dopamine flows from the imperatives of evolution and survival. Dopamine encourages us to maximize our resources by rewarding us when we do so—the act of doing something well, of making our future a better, safer place, gives us a little dopamine "buzz."

TENACITY

I have not failed. I've just found 10,000 ways that won't work.
—Thomas A. Edison

A young man who had recently graduated from college came to see a mental health specialist because he found himself unable to navigate his new world. He hadn't distinguished himself at school, but he had gotten by and managed to graduate in the usual four years. He believed that the structure of school and the built-in pressure to get things done on time had helped keep him on track. Now he was lost.

He didn't have a job, and he didn't know what he wanted to do. The only thing that interested him was smoking marijuana. He had a job waiting tables for a little while but got fired for showing up late or skipping work entirely. His father got him an office job, but he lost that as well because it was obvious to everyone in the office that he had no interest in the work he was given. He was careless and bored, and eventually people just avoided him.

It was the same with relationships. While he was in college he had a long-term relationship with a young woman, but after graduation she broke up with him. His therapist thought that was a good thing because she had exploited him, making him buy her gifts and asking him to do all sorts of chores while showing no signs of affection. The young man knew she didn't care about him, but he kept going back anyway, hoping to restart the relationship. She refused, but continued to take advantage of him in whatever way she could; for example, asking him to drive four hours to bring her a table lamp she wanted for her apartment.

The therapy was a failure. Therapy is hard work, and this young man didn't have it in him. He tried four different therapists who used a variety of techniques, but nothing changed. Three years later he still didn't know what he wanted to do with his life, still smoked marijuana, and was still trying to get back together with his old girlfriend.

The world doesn't always work the way we expect it to. We learn at an early age that Scotch tape does a great job fixing tears in paper, but it doesn't do so well with broken toys and smashed-up dinner plates. The entrepreneur who develops the next killer technology in his garage is often surprised to find that the world isn't beating a path to his door. Success takes years of hard work and so many revisions to the original idea that it's barely recognizable by the time it gets to market. It's not enough to just imagine the future. To bring an idea to fruition we must struggle with the uncompromising realities of the physical world. We need not only knowledge but also tenacity. Dopamine, the chemical of future success, is there to deliver.

THE CASE OF THE RESOLUTE RATS

One way to study tenacity in a laboratory is to measure how hard a rat will work to get food, typically by counting the number of times it will press a lever that sends a food pellet sliding down a chute into its cage. By increasing the number of lever presses required to get the food, scientists can find out whether their rats have the determination to increase their efforts accordingly.

Researchers from the University of Connecticut wanted to see if they could manipulate a rat's tenacity by changing the activity of dopamine in its brain. They put a cage full of rats on a reduced-calorie diet until the animals lost 15 percent of their weight—for comparison, that's like a typical adult losing about 25 pounds. After the rats were good and hungry, the scientists gave them an opportunity to work for rewards in the form of Bioserve tablets, delicious treats (to rats, at least) that come in a variety of flavors, including chocolate marshmallow, piña colada, and bacon.

They began by dividing the rats into two groups. They designated the first group as the control group, and did nothing to them beyond the diet. As for the second group, the scientists injected a neurotoxin into their brains that destroyed some of their dopamine cells. Then they began the experiment.

The first experiment was easy. To receive a Bioserve treat, each rat had to press the lever only one time. Since essentially no work was needed—no tenacity required—this experiment established a necessary condition: it demonstrated that dopamine-deficient rats liked the treats as much as normal rats. This was important, because if dopamine-deficient rats no longer wanted Bioserve goodies, the scientists would not be able to test how hard they would work for them.

When no work was required, the dopamine-deprived rats pressed the lever as many times as the normal rats, and devoured the treats they had earned. This outcome was not surprising because liking and enjoying would not be expected to change as a result of a dopamine alteration. Things did change, though, when the rats had to work harder:

> When the required number of lever presses was increased from one to four, the normal rats pressed their levers nearly a thousand times over the course of 30 minutes. The dopamine-depleted rats weren't as motivated; they pressed the lever only about six hundred times.

> When the requirement was increased to sixteen presses, the normal rats produced nearly two thousand presses, while the dopamine-depleted rats barely increased their presses at all. They were getting only one-quarter the number of treats, but they wouldn't work harder.

> Finally, the requirement was bumped all the way up to sixty-four presses for a single Bioserve tablet. The normal rats managed about twenty-five hundred presses—more than one press per second for the entire 30 minutes. The dopamine-depleted rats didn't increase their work at all. In fact, they pressed less than they had before. They simply gave up.

Removing dopamine appeared to diminish a rat's will to work. But one more experiment was done to confirm that it was tenacity that was affected by dopamine destruction, not liking.

Ice cream is always nice, but if you've just finished a big meal, you probably won't want as much dessert as you would if you hadn't eaten. How much ice cream you want has nothing to do with whether you're hard-working or lazy. It's just that food doesn't mean as much when you're not hungry. So the scientists added a new dimension to the experiment: they manipulated hunger.

The scientists brought in a new group of rats, gave them a good meal, then put them through the experiment. At all levels of effort—even one single press—the pre-fed rats pressed the lever half as much as the hungry ones. When the requirement was doubled, they doubled their efforts. When the requirement was quadrupled, they quadrupled their efforts. But they always stopped at about one-half the presses of the hungry rats. They didn't slack off. They didn't give up. They just didn't want to eat as many pellets because they weren't hungry.

The results reveal a subtle but vital distinction. The feeling of hunger (or the absence of hunger) changed how much the rats valued the pellets, but it did not diminish their willingness to work. Hunger is an H&N phenomenon, an immediate experience, not an anticipatory, dopamine-driven one. Manipulate hunger, or some other sensory experience, and you affect the *value* of the reward earned through work. But it's dopamine that makes the work possible at all: no dopamine, no effort.

This points us toward an understanding of how dopamine affects the choices we make between working hard or taking the easy way. Sometimes we want a fancy meal, and we're willing to work hard to prepare it. Other times all we want to do is "veg out"—we'll tear open a bag of Cheetos in front of the TV, instead of working for even the few minutes it might take to make a simple meal. Consequently, the next step in the experiments was to introduce the element of choice.

The scientists set up a cage with a Bioserve machine and a bowl of lab chow. The lab chow was bland but freely available—no work required. To get the much tastier Bioserve tablets, a rat would have to make four lever presses—minimal effort, but effort nonetheless. The rats with normal dopamine went right for the Bioserve treats. They were willing to do a little bit of work to get something better. The dopamine-depleted rats, on the other hand, headed over to the easy-access lab chow.

The ability to put forth effort is dopaminergic. The quality of that effort can be influenced by any number of other factors, but without dopamine, there is no effort at all.

SELF-EFFICACY: DOPAMINE AND THE POWER OF CONFIDENCE

A bacon-flavored Bioserve treat may be all it takes to motivate a rat, but humans are more complicated. We need to *believe* we can succeed before we are *able* to succeed. This influences tenacity. We have greater tenacity when we encounter early success. Some weight-loss programs help you lose six or seven pounds in the first few weeks. They plan it this way because they know that if you begin with no more than a pound or two loss in that time, you are likely to drop out. They know you are more likely to stick with it if you see that you are capable of doing it. Scientists call this *self-efficacy*.

Drugs such as cocaine and amphetamine boost dopamine, and one result is an increase in self-efficacy, often to pathological levels. People who abuse these drugs may confidently take on so many projects that it is impossible to complete them all. Heavy users may even develop grandiose delusions. With no evidence whatsoever they may believe they will write the most brilliant treatise ever produced, or invent a device that will solve the world's problems.

Under normal circumstances, robust self-efficacy is a valuable asset. Sometimes it can act like a self-fulfilling prophecy. Having a confident expectation of success can make obstacles melt before your eyes.

DOMINATION IN A PILL: SIDE EFFECTS INCLUDE OPTIMISM, WEIGHT LOSS, AND DEATH

In the early 1960s, doctors prescribed large amounts of dopamine-boosting amphetamine to promote "cheerfulness, mental alertness, and optimism," as described by a contemporary advertisement. Most of these prescriptions were written for women, who were twice as likely as men to be prescribed amphetamine to "adjust their mental state." As one doctor described it, amphetamine allowed them to be "not only capable of performing their duties, but to actually enjoy them." In other words, if you don't like cooking or cleaning, it helps to be on speed.

But that's not all. In addition to making housewives happy and productive, it also kept them thin. According to *Life* magazine, two billion tablets were prescribed annually in the 1960s for this purpose alone. But although people did lose weight, it was only temporary, and often at a high cost. Stop using the drug, and the weight comes right back. Keep using the drug and tolerance develops, so the user must take higher and higher doses to get the same effect. That's dangerous. Too much amphetamine can bring about personality changes. It can also cause psychosis, heart attacks, strokes, and death.

"I felt charming, witty, and clever, talking to everybody," wrote one amphetamine user. "I felt a compulsion to make subtle, condescending comments to the more-dimwitted customers [at work] under the guise of being straightforward and helpful. My family has also told me that I've become much more arrogant, snide, and condescending, and my brother tells me that I've been thinking I'm 'hot shit' lately, but he might be jealous of me." Another user said simply, "I used to feel like a young god on speed." The difference is that young gods don't suffer side effects that kill.

A college student needed to get to the airport to fly home for spring break. As with most college students, money was tight, so she made a reservation with a shuttle service that would take her to the airport for only $15. The shuttle had a regular schedule of stops, and she arranged to be picked up at a nearby hotel at 12:30 PM.

She didn't start to get nervous until 1:00. When 1:30 rolled around and there was still no car, she knew something was wrong. By 2:00 she was beginning to sweat. She had called the service repeatedly, and each time she had been assured that "the driver is on his way." She had declined the doorman's friendly offer to call her a taxi, but now she was running out of time.

Thirty minutes and $40 later she stepped out of a taxi at the airport and headed straight to the shuttle reservation desk. She demanded that they reimburse her for the difference between the shuttle and the taxi. It was clearly their fault. They had promised to pick her up at 12:30 and they had failed to keep their promise. It wasn't fair that she should have to pay the difference. It was a matter of justice. The clerk at the reservation desk had no authority to give her the money, but the woman was so sure she was right, that it was inconceivable to her that she would not prevail. It didn't take long before the clerk opened the cash register, and handed over $25.

How does this work? How does a confident expectation of success cause others to give way, even when it seems like it's not in their interest to do so? It's usually because of things that are happening outside of their conscious awareness.

Researchers from the Graduate School of Business at Stanford University wanted to know how subtle, nonverbal behavior affected people's perceptions of one another. They noted that when people expand themselves, taking up a large amount of space, they're perceived as dominant. Conversely, when they constrict themselves, taking up as little space as possible, they're perceived as submissive.

They designed a study to explore the effects of nonverbal displays of dominance or submission. The researchers put two people of the same sex in a room and asked them to discuss photos of famous paintings. They did this in order to conceal the true nature of the study. Only one of the people was an actual test participant. The other was a confederate, secretly working for the researchers. The confederate assumed either a dominant posture (one arm draped over the back of an empty chair next to her, legs crossed so her right ankle rested on her left thigh), or a submissive posture (legs together, hands in her lap, hunched slightly forward). The question was, would the participant mirror the confederate's posture or adopt a complementary, opposite posture?

Most of the time, we mirror the actions of people we're talking to. If one person touches his face or gestures with his hands, so does the other. But this time it was different. When it comes to dominant and submissive postures, the research participants were more likely to adopt a complementary posture rather than mirror the same posture. Dominance triggered submission, and submission triggered dominance.

It didn't happen all the time, though. A minority of participants mirrored the confederate. Would that have an effect on the underlying relationship? The researchers gave the participants a survey to fill out. They wanted to know how they experienced the interaction with the confederate. Did they like her? Did they feel comfortable with her? It didn't matter if the confederate took a dominant or submissive posture. Participants who took the complementary posture not only liked the confederates more, they also felt more comfortable with them compared to the participants who mirrored the confederates.

Finally, researchers asked the participants a series of questions to find out if they were aware of how they were responding to the confederate. Did they know their posture was being influenced by the posture of the other person in the room? It turned out they had no idea. It all occurred outside of their consciousness.

We unconsciously know when someone has a high expectation of success, and we get out of their way. We submit to their will—the overwhelming expression of their self-efficacy, powered by control dopamine. Our brains evolved this way for a good reason: it's a bad idea to get into fights you can't win. If you're picking up signals that your adversary has a high expectation of success, the odds are that this is a fight you want to avoid. This type of behavior is clearly seen in non-human primates. Chimpanzees observing a dominant display constrict themselves to appear as small as possible. On the other hand, when chimps respond to dominant displays with mirrored dominant displays, it usually marks the beginning of a long period of conflict that often ends in violence.

ON ANY GIVEN SUNDAY

Sports lore is rife with stories about underdogs: the phenom overcoming hardscrabble roots, the plucky second-stringers who win the championship, the walk-ons who make it to the pros—in short, the come-from-behind victory over another player, another team, or life itself. Sports movies are almost exclusively about this kind of thing: *Remember the Titans, Rudy, The Bad News Bears, A League of Their Own, Rocky, Hoop Dreams, The Karate Kid*. But the question remains: How does a player or a team demonstrably inferior in skill and ability prevail over a superior opponent? It happens too often to attribute it only to luck. The answer is self-efficacy. One of the most dramatic examples of self-efficacy in sports took place on January 3, 1993, in an NFL playoff game fans call simply "The Comeback."

In the third quarter, the Buffalo Bills were down 35-3 against the Houston Oilers. Bills fans were filing through the exits as a Houston radio announcer commented that although the lights had been on in the stadium since morning, "you could pretty much turn them out on the Bills right now."

But as the clock wound down, things began to change. Luck played some role—a bad kick, a dubious call that went in the Bills' favor—but even that does not account for the burst of success the team experienced. As their comeback began, the Bills scored 21 points in 10 minutes. A player recalled later, "We were scoring at will." As the Oilers proved unable to stop them, a Bills player on the sidelines began to shout, "They don't want it! They don't want it!" Buffalo's will—their belief that they were destined to prevail, their *self-efficacy*—was stronger that day than their opponents' skills and abilities. The Bills sent the game into overtime and won on a 32-yard field goal, 41–38. This victory would be the greatest point-deficit comeback in NFL history.

Of note: Bills star quarterback Jim Kelly had been injured the previous week and was replaced in the Oilers game by his backup, Frank Reich. At that time, Reich held the record for the biggest comeback in college football history. A decade earlier he had led the Maryland Terrapins from a first-half deficit of 31–0 to a 42–40 win over the undefeated Miami Hurricanes. Four years after the Bills' victory over the Oilers, the team, led by quarterback Todd Collins, would come back from a 26-point deficit to defeat the Indianapolis Colts, setting the second-highest point record for a regular-season comeback. The self-efficacy of the Buffalo Bills seemed to propagate itself. Success inspired confidence; confidence produced success.

WHAT IF YOU TRIED BEING NICE?

James was referred to treatment by his employer after he threw a stapler across the room in a fit of rage. He was a middle-aged man who had risen through the ranks to become a vice president at a large company. He was not liked and the only reason for his success was his determination and hard work. He told the therapist that he would have been fired long ago if he had not made himself such a valuable asset. The problem was that he was always angry.

He had been abused as a child and had never come to terms with what had happened. He never told anyone about it, and persuaded himself that it didn't matter because it had happened so long ago. He had been divorced twice, and by this time he had given up on relationships, devoting himself entirely to his work.

His anger had become progressively worse over the years. On one occasion he was ejected from a grocery store for screaming obscenities at a woman who had bumped his shopping cart, and on another he was arrested after shoving a taxi driver during a disagreement over the fare. The charges had been dropped, and James maintained he had been fully justified in what he did. Now, however, he was worried. His job meant everything to him, and he was willing to do anything to keep it, even confront his past.

James had little emotional resiliency and his therapist worried that digging into the trauma would activate disturbing emotions and make his behavior worse before it got better. So before they began to explore the past, they talked about ways to make the present a little less stressful. The therapist wanted to find a way to reduce the constant conflict James had with pretty much everyone he met. So she taught James to be manipulative.

It would be a long time before James could trust anyone, but he wasn't stupid. He quickly learned that he could get his way more easily by smiling at people instead of glaring at them. He began to greet his coworkers in the morning, not because he cared about them, but because it made it easier to get them to finish projects on time. He ordered pizza for his team when they had to work late and complimented people on their appearance. He became a master manipulator.

And he enjoyed it. He liked the new source of power he had found, but he also liked the smiles he got back. A turning point occurred when one of the administrative assistants burst into his office in tears, telling him that someone had opened a credit card account in her name and now she was being threatened by a collections agency. She had chosen him for comfort and advice. Later that week he and his therapist began to talk about his past.

So far we have focused on domination as primarily a solo pursuit, but we cannot achieve every goal by ourselves. Consider domination that requires working with other people.

A relationship that is formed for the purpose of accomplishing a goal is called *agentic*, and it is orchestrated by dopamine. The other person acts as an extension of you, an agent who assists you in achieving your goal. For example, relationships we make at networking events are primarily agentic, and typically result in mutual gain. *Affiliative* relationships, on the other hand, are for the purpose of enjoying social interactions. The simple pleasure of being with another person, experienced in the here and now, is associated with H&N neurotransmitters such as oxytocin, vasopressin, endorphin, and endocannabinoids.

Most relationships have both affiliative and agentic elements. Friends who like to hang out together in the here and now (affiliative) may also work on future projects together, such as planning a white-water rafting trip or an evening at the clubs (agentic). Coworkers with primarily agentic relationships usually enjoy each other's company. Some people are more comfortable in agentic relationships because they're more structured, while others prefer affiliative relationships because they find them more fun. Some people are comfortable with both, others with neither.

There are personality types for each variety of relationship preference. Agentic people tend to be cool and distant. Affiliative people are affectionate and warm. They are also social, and turn to others for support. People who are good at both affiliative and agentic relationships are friendly, accessible leaders, such as Bill Clinton or Ronald Reagan. Those who are less able to navigate agentic relationships are more likely to be friendly, accessible followers. Those who have trouble with affiliative relationships but who are skilled with agentic ones may be viewed as cold and uncaring, whereas those who are poor at both come across as aloof and isolated.

Agentic relationships are established for the purpose of dominating one's environment to extract as much as possible from the available resources, the domain of control dopamine. Although we think of domination as an active, even aggressive, activity, it doesn't have to be. Dopamine doesn't care how something is obtained. It just wants to get what it wants. So an agentic relationship can be entirely passive; for example, when a manager running an employee meeting gets the outcome he wants by keeping quiet.

Agentic relationships can easily become exploitative, such as when a scientist enrolls participants in a dangerous experiment without telling them the risks, or when an employer hires someone under false pretenses to exploit her hard work. But an agentic relationship can be beautifully humane, too. Ralph Waldo Emerson, the American poet, wrote: "Shall I tell you the secret of the true scholar? It is this: Every man I meet is my master in some point, and in that I learn of him."

No matter how ignorant, degraded, or foolish a man is, there is something he knows, something he has mastered, that Emerson valued. Emerson sought to find intellectual worth in all people, regardless of their station in life. Such a relationship is agentic because the relationship is about gain—gaining knowledge. It's not about the H&N pleasure of having company. What makes this dopaminergic quotation particularly interesting is that Emerson called this man "my master." He wrote of domination through submission—self-submission in the form of deference, humility, and obedience.

SUBMISSIVE MONKEYS, HUMBLE SPIES

When researchers at the Illinois State Psychiatric Institute injected a dopamine-boosting drug into stump-tailed macaque monkeys, they observed an increase in submissive gestures, such as lip smacking, grimacing (the monkey version of smiling), and holding out an arm to another monkey for a gentle bite. On the surface this response doesn't make sense. Why would dopamine, the neurotransmitter of dominance, trigger submissive behavior? Is there a contradiction here? Not at all. In the control circuit, dopamine drives domination of the environment, not necessarily the people in it. Dopamine wants more, and it doesn't care how it gets it. Moral or immoral, dominant or submissive, it's all the same to dopamine, as long as it leads to a better future.

Consider a spy stationed in a hostile country, trying to gain access to a government building. While prowling around a back alley, he runs into the janitor. The spy treats the janitor as his equal, perhaps even his superior, in order to gain his cooperation—submissive behavior aimed at dominating the environment and reaching his goal.

Submissive behavior can have negative connotations—letting people "walk all over you," for instance—but the scope of submissive behavior is much wider than that. In modern society, submissive behavior is often a sign of elevated social status—think of the strict adherence to manners, the focus on social customs, and, in conversation, the deference to others that is part and parcel of the behavior of what we might call "the elite." The common name for this behavior is *courtesy*, a word derived from the word *court*, because it was the behavior originally adopted by the nobility. By contrast, dominant behavior, representing the opposite of courtesy, may stem from personal insecurity or an imperfect education.

Planning, tenacity, and force of will through personal effort or by working with others: these are the ways control-circuit dopamine lets us dominate our environment. But how do we behave—and feel—when the system falls out of balance? In particular, what happens when there is too much or too little control dopamine?

OUTER SPACE CHALLENGE, INNER SPACE STRUGGLE

GQ Magazine: What does it feel like to go to the moon?
Buzz Aldrin: Look, we didn't know what we were feeling. We weren't feeling.

GQ: What were your emotions as you walked on the surface of the moon?
BA: Fighter pilots don't have emotions.

GQ: But you're a human!
BA: We had ice in our veins.

GQ: Well, did you ever say, "I'm going to get in that [fragile lunar module], and land on the moon"? Did that ever sort of flabbergast you?

BA: I understood the construction of it. It's got landing gear. It's got struts that compress. It's got probes that hang down. It was a marvel of engineering.

—Interview with Buzz Aldrin

Instead of taking a bow for walking on the moon, Colonel Buzz Aldrin, PhD, told his admirers, "It's something we did. Now we should do something else," apparently no more satisfied than if he had painted a fence. His desire was not to bask in his glory but to find "something else"—the next big challenge that could hold his interest. This perpetual need to identify a goal and calculate a way to reach it was perhaps the most important factor in his historic success. But it's not easy having so much dopamine coursing through the control circuits. It almost certainly played a significant role in Aldrin's post-lunar struggle with depression, alcoholism, three divorces, suicidal impulses, and a stay on a psychiatric ward, which he described in his candid autobiography, *Magnificent Desolation: The Long Journey Home from the Moon.*

Just as desire dopamine facilitates becoming addicted to drugs—chasing the high and receiving less and less dopamine "buzz" from it—some people have so much control dopamine that they become addicted to achievement, but are unable to experience H&N fulfillment. Think of people you know who work relentlessly toward their goals but never stop to enjoy the fruits of their achievements. They don't even brag about them. They achieve something, then move on to the next thing. One woman described taking a leadership position in a division of a company that was in chaos. Years of long hours and hard struggle allowed her to get everything running smoothly, and she immediately became bored. For a few months she tried to enjoy the new, relaxed environment she had created, but she couldn't bear it, and requested a transfer to a department that was a complete mess.

These individuals exhibit the effects of an imbalance between future-focused dopamine and present-focused H&N neurotransmitters. They flee the emotional and sensory experiences of the present. For them, life is about the future, about improvement, about innovation.

Despite the money and even fame that comes from their efforts, they are usually unhappy. No matter how much they do, it's never enough. The family crest of James Bond, the resourceful, relentless, often ruthless secret agent, contains the motto *Orbit Non Sufficit:* The World Is Not Enough.

Colonel Aldrin faced this problem in a more profound way than perhaps any human being ever had: *I have walked on the surface of the moon. What could I possibly do to top that?*

DOPAMINE EXPLAINS THE MYSTERIES OF ADHD

What about people on the other end of the spectrum, people whose control dopamine circuits are weak? Their struggle with internal control manifests itself as impulsivity and difficulty keeping themselves focused on complex tasks. This problem can result in a familiar condition: attention deficit hyperactivity disorder (ADHD).[2] Poor focus, concentration, and impulse control can severely interfere with their lives, and it can make them difficult to be with. Sometimes they don't pay attention to details, or follow through on tasks. They may start paying bills, then switch to doing the laundry, then change a light bulb, then sit down and watch TV with everything strewn all over the place. During conversations, they can become distracted easily, and not listen to what people say to them. Sometimes they don't keep track of time, making them late, and they may lose things, such as car keys, cell phones, even passports.

ADHD is seen most often in children, and for good reason. The frontal lobes, where control dopamine acts, develop last, and do not fully connect to the rest of the brain until a person finishes adolescence and enters adulthood. One of the jobs of the control circuit is to keep the desire circuit in check; hence the impulse control problem

---

2   This illness is commonly called *attention deficit disorder*, or *ADD*, because adults usually don't have the hyperactivity seen in children. Nevertheless, we'll use the scientific term, ADHD.

associated with ADHD. When control dopamine is weak, people go after things they want with little thought about the long-term consequences. Kids with ADHD grab toys and cut in line. Adults with ADHD make impulse purchases and interrupt people.

The most common treatments for ADHD are Ritalin and amphetamine, stimulants that boost dopamine in the brain. When these drugs are used to treat people with ADHD, tolerance usually doesn't develop as it does for those who take these drugs to lose weight, get high, or enhance their performance. Nevertheless, stimulants are addictive drugs. The FDA puts them in the same class as opioids, such as morphine and OxyContin. These are considered the highest risk in terms of abuse, and they have the most stringent restrictions on how doctors can prescribe them.

People who live with ADHD are at high risk of addiction, especially adolescents, because of their poorly functioning frontal lobes. Years ago, when the illness was less well understood, doctors and parents were reluctant to give these vulnerable children addictive drugs such as Ritalin and amphetamine. It sounded reasonable: don't give addictive substances to people at risk for addiction. But rigorous testing showed unambiguously that adolescents who were treated with stimulant drugs were less likely to develop addictions. In fact, those who started the drug at the youngest age and took the highest doses were the least likely to develop problems with illicit drugs. Here's why: if you strengthen the dopamine control circuit, it's a lot easier to make wise decisions. On the other hand, if effective treatment is withheld, the weakness of the control circuit is not corrected. The desire circuit acts unopposed, increasing the likelihood of high-risk, pleasure-seeking behavior.

A SURPRISING RISK AMONG ADHD PATIENTS

Drug addiction isn't the only risk these children face. It's hard for a child with ADHD to extract valuable resources from his environment—typically in the form of good grades—when he can't focus or control his impulses. But poor grades are only the beginning. Young people with ADHD have difficulty making friends. Who wants to be around someone who interrupts, grabs things, and doesn't wait their turn? They often have to read homework assignments over and over again before they understand the material. This happens as a result of constant distractions. Spending that much time on homework doesn't leave much time for extracurricular activities, such as sports and clubs. With few friends, poor grades, and cut off from healthy sources of pleasure, children living with untreated ADHD become more willing to pursue unhealthy sources of pleasure. In addition to drugs, they may also have problems with early sexual activity and overeating, particularly "pleasure foods" that are high in salt, fat, and sugar.

A massive study involving 700,000 children and adults, including 48,000 with ADHD, found that children with ADHD were 40 percent more likely to be obese, and adults were 70 percent more likely to be obese. At nearly three-quarters of a million participants, with data taken from cultures around the world, the study was not only greater in size than most investigations of its type but also far more diverse, allowing the scientists to compare the results from different countries where one finds a variety of diets and eating rituals. Yet, in spite of the differences in diets among, for instance, Qatar, Taiwan, and Finland, the findings were the same. Country of residence did not affect the relationship between ADHD and obesity. There was also no difference between men and women.

Despite the strengths of this study, there are weaknesses as well. Just because we find that people with ADHD are more likely to be obese doesn't necessarily mean that having ADHD *causes* obesity. What if it was the other way around? What if being overweight somehow affected the brain in a way that caused ADHD? The fancy scientific term way of saying this is *association does not imply causation*. Just because two things are found together doesn't necessarily mean that one caused the other.

We'd have more confidence that ADHD leads to obesity if we could show that people develop symptoms of ADHD *before* they become obese. So researchers from the Universities of Chicago and Pittsburgh evaluated nearly 2,500 girls to find out if there was a connection between unhealthy weight and problems with impulsivity. The lead researcher noted, "Children are constantly cued to eat by food commercials, vending machines, etc., so it is easy to imagine how a child who is poorly inhibited could have difficulty resisting these cues to eat."

The results were as expected. Girls who had problems with impulsivity and planning at age 10 gained more weight over the following six years. The scientists reported that a significant amount of the weight these girls gained came from bingeing—intense bursts of no self-control.

For a similar reason, overweight children are more likely to be hit by cars when they're crossing the street. It's not because they walk more slowly; it's because they're impulsive. Researchers at the University of Iowa collected 240 children who were seven or eight years old, and asked them to cross a busy street to measure how long they waited and how often a child was hit by a car.[3]

Although overweight people sometimes walk more slowly, in this experiment weight had no effect on how fast the children crossed the street. But there was a direct relationship between how overweight the child was and how quickly he or she stepped out into traffic. Less overweight children waited longer than more overweight children. Overweight children also left a smaller buffer between themselves and oncoming traffic—that is, they allowed the cars to get closer. Not surprisingly, they were hit more frequently.

It's important to remember that biology is not destiny. People whose control-dopamine systems are at one extreme or the other can change. People with ADHD can improve dramatically with medication, psychotherapy, and sometimes just time. Colonel Aldrin, who faced a different problem, eventually found ways to harness the intensity of his creative drive. Since returning from the moon, he has written or cowritten a dozen books, produced a computer strategy game, and proposed a revolutionary method of space travel that could make a crewed mission to Mars more practical. He also found time to appear on numerous TV shows, including *Dancing with the Stars*, *The Price Is Right*, *Top Chef*, and *The Big Bang Theory*.

---

3   Nobody really got hit by a car. The researchers used virtual reality.

THE CHEMISTRY OF FRAUD

I know your noble nature hates the thought of treachery
or fraud. But what a glorious prize is victory!
—Sophocles, *Philoctetes*

I like to win, but more than anything, I can't stand this
idea of losing. Because to me, losing means death.
—Lance Armstrong

In 1999, after surviving a battle with advanced cancer, Lance Armstrong won his first Tour de France. A reporter for the New York Times *described him in a way that would come to be typical in the following years: "a man of strong will and focus" who "dominated the Tour." He went on to win seven consecutive Tour de France races, dominating not only that famous race, but the sport itself.*

Armstrong was legendary for his determination. He preferred to bike with a headwind because it made the course harder and gave him more opportunities to outlast the competition. Author Juliet Macur described Armstrong's determination with this story: "[A tree] was once on the other side of [his] property, 50 yards west of his house. Armstrong wanted it at the front steps. The transplantation cost $200,000. His close friends joke that Armstrong, who is agnostic, engineered the project to prove he didn't need God to move heaven and earth."

"I think I would probably go crazy if I was 35 or 40 and didn't have some competition in my life," Armstrong said.

In 2012, the world-champion cyclist was stripped of all seven of his Tour de France titles when it was revealed that he had used performance-enhancing drugs. Why would this legendary athlete cheat, this man of steely determination who never gave up, even in the face of cancer? Oddly enough, he may have cheated because he was so successful.

Dopamine doesn't come equipped with a conscience. Rather, it is a source of cunning fed by desire. When it's revved up, it suppresses feelings of guilt, which is an H&N emotion. It is capable of inspiring honorable effort, but also deceit and even violence in pursuit of the things it wants.

Dopamine pursues *more*, not *morality*; to dopamine, force and fraud are nothing more than tools.

Israeli researchers designed an experiment to help them better understand why people cheat. They set up a pair of games that would pit one player against another. The first was a guessing game in which players competed to see who could guess what images were going to appear on a computer screen. In this game it was impossible to cheat. The second game was different: the first player rolled a pair of dice, and announced the results to the second player. The higher the roll, the more money the first player got, and the less her opponent got. In this game cheating was not only possible, it was easy. The second player couldn't see the actual dice, so the first player could report anything she liked. The winner and the loser of the first game took turns rolling the dice and announcing the result.

Because of the way dice are marked, if everyone was honest, the average score should have been about seven. The losers of the first game reported an average roll of a little over six during the second game, which was consistent with random chance. The winners of the first game, on the other hand, reported a second-game average of almost nine. Statistical analysis revealed that it was extremely unlikely that number could have come about by chance. There was a greater than 99 percent likelihood that the first-game winners cheated on the second game.

For the next phase of the experiment, the researchers changed things. Instead of a competition, the first game was changed to a lottery—and the new arrangement yielded a very different outcome. The players who won the lottery didn't cheat at all on the second game. In fact, they appear to have underreported their scores, resulting in their opponents sharing the spoils of victory.

The scientists weren't sure how to explain this result. They thought that maybe people who won competitions, as opposed to winning by pure luck, developed a sense of entitlement that allowed them to justify subsequent cheating. But by thinking about the role dopamine plays in motivating us to dominate our environment, we can find a better explanation.

Winning competitions, along with eating and having sex, is essential for evolutionary success. In fact, it's winning competitions that gives us access to food and reproductive partners. As a result, it's not surprising that winning competitions releases dopamine. It's the rush of pleasure we feel when we send the tennis ball flying over the net, get a good grade on a test, or receive praise from our boss. The surge of dopamine feels good, but it's different from a surge of H&N pleasure, which is a surge of satisfaction. And that difference is key: the dopamine surge triggered by winning leaves us wanting more.

WINNING TO KEEP FROM LOSING

It's not enough to win the Tour de France. It's not enough to win it twice or even seven times. Winning is never enough. *Nothing* is ever enough for dopamine. It is the pursuit that matters, and the victory, but there is no finish line, and never will be. Winning, like drugs, can be addictive.

Yet the pleasurable rush that never satisfies is only half of the equation. The other half is the dopamine crash that feels so awful.

Every year, physicians in Washington, DC, fill out a ballot in which they vote for the best doctors in a variety of medical specialties. The results are published in the *Washingtonian* magazine's famous Top Doc issue. It's their best-selling issue. Being named a Top Doc is an honor, and it feels nice. Your colleagues see it, your friends and family see it, everybody sees it. After the glow of satisfaction wears off, though, an uncomfortable question comes up: *Will I make it next year? All the people who congratulated me—what will they think when my name disappears from the list? No one stays on the list forever; how will I bear the humiliation of being dropped?* No one likes to lose, but it's ten times as bad after you win. When you open the magazine expecting to see your name and it's not there, you get an unpleasant feeling in the pit of your stomach.

Winners cheat for the same reason that drug addicts take drugs. The rush feels great, and withdrawal feels terrible. Both know that their behavior has the potential to destroy their lives, but the desire circuit doesn't care. It only wants more. More drugs, more success. But true success doesn't come from cheating. If you make a mistake, people will forgive you, but if you act dishonestly, it will stick with you for a long time. That's why the control circuit is so important. It's rational. It's able to make cool, reasoned decisions, ones that will maximize your welfare not just today, but far into the future. And yet, for many people, fraud is a powerful, sometimes overwhelming temptation when chasing the high of victory. At least in the short term, fraud works.

Or you could just punch somebody.

HOT AND COLD VIOLENCE

Dr. Jones stood in the elevator, dreading the patient interview that was about to happen. It was 1:00 AM, and she had been called down to the emergency room to evaluate a patient who said he was going to kill someone. She had to get it right. When a psychiatric patient follows through on a threat to commit murder, the victim dies, the killer must be caught—and the doctor who set the killer free can be held responsible.

The patient, disheveled and malodorous, stared unblinking at Dr. Jones. He had been here before. He had been disruptive and uncooperative. During one stay he was accused of inappropriately touching a woman being treated for schizophrenia. He claimed he was allergic to all psychiatric medications except Xanax.

Apart from cocaine use there wasn't much wrong with him psychiatrically, but tonight he demanded to be admitted to the hospital. He mentioned multiple arrests and a three-year stint in prison. If he wasn't taken up to "the unit," he said, he would carry out his plan and kill someone.

"Let's just say it's someone who did something to me, okay?" he said.

Paranoia is one of the most treatable psychiatric conditions associated with people who threaten violence. Paranoia makes them feel afraid, and sometimes they conclude that the only way to protect themselves is to kill the people they imagine are plotting against them. With antipsychotic

medications, the delusions, along with the risk of violence, usually go away
in about a week.

But the patient who sat across from Dr. Jones, with his eyes still drill-
ing into hers, wasn't psychotic.

Dr. Jones faced a dilemma. She knew that the patient wouldn't ben-
efit from an inpatient stay, and admitting him to the unit would put other
patients at risk. On the other hand, he had a history of violence. She admit-
ted him, fearing for the safety of the victim he refused to name, but feeling
guilty for potentially endangering the patients on the ward.

Violence is sometimes the result of dysfunction or pathology. But most
of the time, violence is a choice—a coercive and calculated way to get the
thing you want.

Force, often expressed as violence, is the ultimate tool of domination,
but is it dopaminergic?

Violence comes in two flavors: planned violence inflicted for a purpose, and spontaneous violence set off by passion. Violence for a purpose, designed to get something the perpetrator desires, might be as
prosaic as mugging someone on the street, or as earth-shattering as
launching a global war. The emphasis in each case is on effective strategy, planned in advance, sometimes in excruciating detail, and always
aimed at gaining resources or control. This is dopamine-driven aggression, and it tends to have a low emotional content. It is cold violence.

Think of dopaminergic calculation and instinctive response as
opposite ends of a seesaw: when one is high, the other is low. The ability to suppress emotions such as fear, anger, or overwhelming desire
provides an advantage in the midst of conflict. Emotion is almost
always a liability that interferes with calculated action. In fact, a common strategy of domination is to stimulate emotional reactions in one's
adversary to interfere with his ability to execute his plans. In sports it
comes in the form of trash talk on the basketball court or at the line of
scrimmage.

Aggression driven by passion is a lashing out at provocation.
This is not a calculated action orchestrated by the dopamine control
circuit—just the opposite. When passion drives aggression in response

to provocation, dopamine is suppressed by the H&N circuits, and people who display this type of aggression usually degrade their future well-being. They can end up injured, arrested, or simply embarrassed. Think of a parent losing his temper at a child's hockey game. Throwing anything from a fit to a punch is not a calculated move but a thoughtless emotional reaction. From dopamine's perspective there is nothing to be gained, no resources to maximize, no advantage to be taken. Emotion overwhelms control dopamine's consideration, caution, and calculation.

Anthony Trollope, an English novelist, contrasted the two approaches to describe a political debate that took place between two of his characters, Daubeny and Gresham, the leaders of opposing parliamentary parties:

> Whereas Mr. Daubeny hit always as hard as he knew how to hit, having premeditated each blow, and weighed its results beforehand, having calculated his power even to the effect of a blow repeated on a wound already given, Mr. Gresham struck right and left and straightforward . . . , and in his fury might have murdered his antagonist before he was aware that he had drawn blood.

Violence can give us domination, but to be successful, it must come from the cold circuits of control dopamine.

 ## WHAT IS A DOPAMINERGIC PERSONALITY?

Some people have more active dopaminergic circuits than others. Researchers have identified a number of genes that contribute to the development of this type of personality. It's important to note that elevated dopamine activity can express itself in different ways. Someone with a highly active desire circuit might be impulsive or difficult to satisfy, constantly seeking more. His counterpart would be someone who is easily satisfied. Instead of downing shots at a noisy nightclub, a less dopaminergic person might prefer to spend the day gardening and then go to bed early.

Alternatively, someone with a highly active control circuit might be cold and calculating, ruthless and devoid of emotion. Her counterpart would be a warm, generous person, who is more interested in nurturing friendships than winning competitions. The brain is complicated, and the way in which activity in one circuit is translated into behavior depends on activity in many other circuits all working together. In addition to these examples, a dopaminergic personality can be expressed in other ways that we'll describe later. These people all have one thing in common, though. They are obsessed with making the future more rewarding at the expense of being able to experience the joys of the present.

SUPPRESSION OF EMOTION

If you can keep your head when all about you
Are losing theirs and blaming it on you . . .
If you can force your heart and nerve and sinew
To serve your turn long after they are gone,
And so hold on when there is nothing in you
Except the Will which says to them: "Hold on!" . . .
Yours is the Earth and everything that's in it.
—Rudyard Kipling, *If*

Emotion is an H&N experience. It's what we feel right here, right now. Emotion is critical to our ability to understand the world, but emotions can sometimes overwhelm us. When that happens, we make less-logical decisions. Fortunately, dopamine's opposition to H&N circuits can turn down the volume on emotion. In complex situations, people who have what we call "a cool head," people who are more dopaminergic, are able to suppress this response, and make more deliberate choices that often work better. One of our evolutionary ancestors, one endowed with a particularly robust dopamine control circuit, might respond to a charging lion by suppressing the urge to panic, and instead of trying to outrun the lion, he calmly picks up a burning stick from his fire to frighten it away. When bold action is required in the midst of chaos, the one who can stay calm, take stock of available resources, and quickly develop a plan of action is the one who will pull through.

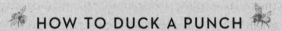 HOW TO DUCK A PUNCH

Although the complexities of modern society can make the automatic decisions of fight-or-flight work against our best interests, in more primitive situations it works just fine. A young doctor talking to an irritable substance abuser in the emergency room found himself unable to comply with the patient's demand for drugs. When it became clear to the patient that he was not going to get what he wanted, he threw a punch. Fortunately, the doctor ducked, and before the patient had time to swing again, help arrived in the form of two security guards who were able to calm the patient down. When it was all over, the doctor said, "I had no idea what was going on. There was no time to think. It just happened." He was pleased to discover that he was the lucky owner of H&N circuits that knew when to duck, no dopamine calculation required.

I took out my 40-foot boat with one crew member, and we sailed toward open ocean. Soon we encountered 35-mile-per-hour winds and 10-foot

*waves. Neither one of us was worried. We had seen this type of weather
many times before.*

*I took the wheel to bring us about. As I was turning, I heard a loud
pop, and the wheel spun freely. I no longer had control of the rudder, and I
was more frightened than I ever remember being in my life.*

*We were within an L-shaped reef. The coral was visible just above the
water, and the waves were pushing us toward it. My first thought was to
jump out of the boat. I wanted to put on a life preserver, and try to swim
out of danger. I quickly realized that would be impossible. The waves
would either smash my body onto the reef, or the undertow would pull me
farther out to sea. I could feel utter panic approaching, and I knew that if I
allowed it to control me, I would lose my ability to think. This all happened
over the course of about 10 seconds.*

*To save myself, I had to start thinking. I radioed a Mayday message,
then my crewman and I got to work on the sails and used them to steer us
out of the reef. We then figured out a way to control the rudder with our feet,
and we got the boat pointed in the direction of the shore. As soon as I began
to plan and act, the panic receded, and I could think rationally.*

*After we made it to shore, while I was walking back to my room, I
began to weep and shake uncontrollably.*

This real-life story is an excellent example of the interplay between
dopamine and the H&N chemical of fight or flight, norepinephrine.
When the steering mechanism broke, norepinephrine kicked in. The
H&N emotion of fear overwhelmed the sailor. He just wanted to get
out of the situation. At first, the initial neurochemical H&N flood dis-
placed his dopaminergic ability to plan. Nevertheless, the fact that he
could sense that panic was on the way, but was able to hold it off, is an
indication that his dopamine system had not shut down completely.

After only a few seconds, control dopamine was fully activated, and
he began to make rational plans. H&N norepinephrine was shut down
and the fear receded, leaving a passionless, cerebral approach to survival. After the crisis was over and he was safely on shore, dopamine
receded, leaving room for H&N to come roaring back, triggering the
shaking and weeping.

Conventional wisdom would attribute his survival at sea to "running on adrenaline." In fact, the opposite was true. He wasn't running on adrenaline; he was running on dopamine. During the intense moments when he saved the boat, dopamine was in control and adrenaline (called norepinephrine when it is inside the brain) was suppressed.

In the eighteenth century, Samuel Johnson summarized the situation like this: "When a man knows he is to be hanged in a fortnight, it concentrates his mind wonderfully." A more recent doctor, Dr. David Caldicott, an emergency room physician at Calvary Hospital in Canberra, Australia, expressed it this way: "Emergency medicine is like flying a plane. Hours of mundanity punctuated by moments of sheer terror. If you're worth your salt, you're not scared, though. Just focused."

IT'S EASIER TO KILL FROM A DISTANCE

In the science fiction classic *Dune*, by Frank Herbert, the hero has to prove he is human by suppressing his animal instinct to act in the here and now. His hand is placed in a diabolical contraption, a black box that creates unimaginable pain. If he pulls his hand out of the box, the old woman administering the test will pierce his neck with a poison needle, and he will die. She tells him, "You've heard of animals chewing off a leg to escape a trap? That's an animal kind of trick. A human would remain in the trap, endure the pain, feigning death that he might kill the trapper and remove a threat to his kind."

Some people are naturally better at suppressing emotion than others. In fact, they're born that way, in part because of the number and nature of their dopamine receptors, molecules in the brain that react when dopamine is released. They differ based on genetics. Researchers measured the density of dopamine receptors (how many there are, and how closely they crowd together) in the brains of a variety of people, and compared the results to tests that measured the person's "emotional detachment."

The detachment test measured traits such as the tendency to avoid sharing personal information and to become involved with other people.

The scientists found a direct relationship between receptor density and personal engagement. High density was associated with a high level of emotional detachment. In a separate study, people who had the highest detachment scores described themselves as "cold, socially aloof, and vindictive in their relationships." By contrast, those with the lowest detachment scores described themselves as "overly nurturing and exploitable."

Most people have personalities that fall somewhere between the highest and lowest scores on the detachment scale. We're neither aloof nor overly nurturing. How we react depends on the circumstances. If we're engaged with the peripersonal—up close, in direct contact, focused on the present moment—H&N circuits are activated, and the warm, emotional aspects of our personality come out. When we're engaged in the extrapersonal—at a distance, thinking abstractly, focused on the future—the rational, emotionless parts of our personality are more likely to be seen. These two different ways of thinking are illustrated by the ethics dilemma called "the trolley problem":

A runaway train hurtles down the tracks toward a group of five workers. If nothing is done, they will all die. It's possible, however, to stop the train by pushing a bystander onto the tracks. His death will slow down the train enough to save the five workers. Would you push the bystander onto the tracks?

In this scenario, most people would be unable to push the bystander onto the tracks—unable to kill a person with their own hands even to save the lives of five other people. The H&N neurotransmitters in play are responsible for generating empathy for others and will overwhelm dopamine's calculated reason in most people. The H&N reaction is so strong in this situation because we're so close, right in the peripersonal zone. We would have to actually put our hands on the victim as we send him to his death. That would be impossible for all but the most detached person.

But since H&N's strongest influence is in the peripersonal space—in the immediate realm of what the five senses tell us—what would happen if we moved back, one step at a time, incrementally diminishing H&N's influence on our decision? Does our willingness—our ability—to trade one life for five increase as we get literally farther away from our victim, as we move out of the H&N peripersonal space into the dopaminergic extrapersonal?

Start by eliminating the H&N sensation of physical contact. Imagine you're standing some distance away watching the scene unfold. There's a switch you can pull that will divert the train from the track with five people on it to a track that will kill only one. Do nothing, and the five will die. Will you throw the switch?

Pull back farther. Imagine you are sitting at a desk in a different city on the other side of the country. The phone rings and a frantic railway worker describes the situation. From your desk you control the path of the train. You can activate a switch and divert the train to a track with only one person on it, or do nothing and allow the train to hit the five people. Will you throw the switch?

Finally, make the situation as abstract as possible: squeeze out all the H&N and make it purely dopaminergic. Imagine that you are a transportation systems engineer, designing the safety features of the railway track. Cameras have been installed by the side of the tracks to provide information about who is standing where. You have the opportunity to write a computer program that will control the switch. The program will use the camera information to choose which track will kill the fewest people. Will you write the software that in the *future* might save five people by killing one?

The scenarios change but the outcomes will be the same: one life is sacrificed so that five can be saved, or five lives are lost to avoid the direct killing of one person. Very few people would put their hands on an innocent person's back and push him to his death. Yet very few people would hesitate to write the software that would manage the track switches in a way that minimizes loss of life. It's almost as if there were two separate minds evaluating the situation. One mind is rational, making decisions based on reason alone. The other is empathic, unable to kill a man, regardless of the big-picture outcome. One seeks to dominate the situation by imposing control to maximize the number of lives saved; the other does not. Whether a person chooses one outcome or the other partly depends on activity within the dopamine circuits.

HARD DECISIONS IN THE REAL WORLD

This problem is more than just theoretical; it confronts developers of self-driving cars. If a fatal crash between two cars is inevitable, what should the self-driving car be programmed to do? Should it swerve in one direction to protect the life of its owner, or should it swerve in the opposite direction, killing its owner, if fewer people in the other car will die? Consumer tip: If you're in the market for a self-driving car, ask the salesperson how it's been programmed.

Another example of the problem was depicted in the 2016 film *Eye in the Sky*. Terrorists in Kenya are preparing two suicide bombers for an attack that will kill as many as two hundred people. There's very little time to stop them. On the other side of the world, the remote pilot of a drone is poised to launch a missile to kill the terrorists. Just before he fires, a young girl sets up a table to sell bread next to the terrorists' house. If the drone pilot does nothing, hundreds will die. But to save those lives, he must kill the little girl along with the terrorists. The film documents the intense debate over which choice to make in this realistic portrayal of the "trolley problem."

Sometimes we act one way: cold, calculating, seeking to dominate the environment for future gain. Sometimes we act another: warm, empathic, sharing what we have for the present joy of making others happy. Dopamine control circuits and H&N circuits work in opposition, creating a balance that allows us to be humane toward others, while safeguarding our own survival. Since balance is essential, the brain often wires circuits in opposition. It works so well that sometimes there is even opposition wired into the same neurotransmitter system. The dopamine system operates in this way, so what happens when dopamine opposes dopamine?

THE RADISHES-AND-COOKIES CHALLENGE

The neurotransmitter dopamine is the source of desire (via the desire circuit) and tenacity (via the control circuit); the passion that points the way and the willpower that gets us there. Usually the two work together, but when desire fixates on things that will bring us harm in the long run—a third piece of cake, an extramarital affair, or an IV injection of heroin—dopaminergic willpower turns around, and does battle with its companion circuit.

Willpower isn't the only tool control dopamine has in its arsenal when it needs to oppose desire. It can also use planning, strategy, and abstraction, such as the ability to imagine the long-term consequences of alternate choices. But when we need to resist harmful urges, willpower is the tool we reach for first. As it turns out, that might not be such a good idea. Willpower can help an alcoholic say no to a drink once, but it's probably not going to work if he has to say no over and over again for months or years. Willpower is like a muscle. It becomes fatigued with use, and after a fairly short period of time, it gives out. One of the best experiments that demonstrated the limits of willpower was the famous radishes-and-cookies study. This study relied on deception. Volunteers were told that they were signing up for a food-tasting study. Here is how one scientist described it:

The laboratory room was carefully set up before participants in the food conditions arrived. Chocolate chip cookies were baked in the room in a small oven, and, as a result, the laboratory was filled with the delicious aroma of fresh chocolate and baking. Two foods were displayed on the table at which the participant was seated. One display consisted of a stack of chocolate chip cookies augmented by some chocolate candies. The other consisted of a bowl of red and white radishes.

When the participants arrived, they were hungry. They had been told to skip a meal before coming to the laboratory. The sight and smell of the freshly baked chocolate chip cookies were very tempting under these conditions. One at a time the participants were ushered into the laboratory where the chocolate chip cookies had just come out of the oven, and they were told to sample two or three cookies or two or three radishes, depending on which group they had been assigned to. Before the participant began to eat, the scientist left the room, reminding the participant that she must only eat the food she was assigned to.

None of the participants assigned to the radishes broke the rules and ate a cookie, but they were obviously tempted. The researchers peeked around a curtain to watch what they did. "Several of them [looked] longingly at the chocolate display and in a few cases even pick[ed] up the cookies to sniff at them."

After about 5 minutes, the scientist returned, and told the participant that the next step in the study was something completely unrelated: it was a test of problem-solving ability. What the participants weren't told was that the problem couldn't be solved. The question was, how long would each participant persevere at this impossible task?

The participants who had been allowed to eat cookies worked on the problem for about 19 minutes. The ones who had been only allowed to eat radishes, those who had to exert self-control to counteract their desire for cookies, persisted at the task for only 8 minutes—less than half the time—before they gave up. The researchers concluded, "Resisting temptation seems to have produced a psychic cost, in the sense that afterward participants were more inclined to give up easily in the face of frustration." If you're on a diet, the more times you resist temptation, the more likely you are to fail the next time around. Willpower is a limited resource.

THE WILLPOWER EXERCISE MACHINE

If willpower is like a muscle, can it be strengthened through exercise? Yes, but it requires some high tech "exercise equipment," the kind of equipment that scientists at the Center for Cognitive Neuroscience at Duke University used to see if they could enhance the part of the brain people use for willpower.

First they made things easy. They paid participants money if they completed a task successfully. It's easy to get motivated when there is an immediate reward. Using a brain scanner, they were able to see activation of the ventral tegmental area of the brain, the place where both the desire and control circuits originate. Next they asked the participants to find ways to motivate themselves. They suggested a number of strategies, such as telling themselves, "You can do it!" They encouraged the participants to be creative and use whatever they thought would be most motivating. Some people imagined coaches encouraging them. Others imagined situations in which their efforts were rewarded. All the while, they lay in the brain scanner, and the scientists watched what happened in the motivation region of their brains. They were surprised at what they saw: nothing happened. Although getting money worked, when participants tried to do it on their own, they failed.

Next, the scientists gave them a little help in the form of biofeedback, which is when a person is provided information on how their body and brain are functioning. This information helps them find effective ways to take control of things that are usually unconscious. The best-known form of biofeedback is for relaxation. A device that measures tiny amounts of sweat is attached to a person's finger. The less sweat, the greater the relaxation. The signal is expressed as a tone, and the user tries to manipulate the tone in the direction of relaxation. It works.

In the motivation experiment, the participants were shown a thermometer with two lines. One showed the current level of activity in the motivation region, and the other represented a higher target they should try to achieve. Now they could see which strategies worked and which ones didn't. After a while, they built up a collection of imagined scenes that effectively boosted motivation activity. These strategies continued to work even when the thermometer was removed. Strengthening willpower was possible, but it required a high-tech window that allowed the test participants to look deep inside their brains.

DOPAMINE VERSUS DOPAMINE

Even though it's possible to strengthen willpower, it's still not the answer to long-term, enduring change. So what does work? That question is of great interest to clinicians who help people struggling to overcome addictions. You can't beat drugs with willpower alone. It takes more than that. There are medications that help with some addictions, but they don't work when they're given alone. They have to be combined with some form of psychotherapy.

The goal of addiction psychotherapy is to pit one part of the brain against another. Part of the dopamine desire circuit becomes malignant in drug addiction, pushing the addict into compulsive, uncontrollable use. It has to be opposed by an equally potent force. We know willpower won't do it. What other resources can be called on to win this fight?

This question has been studied extensively, and the knowledge gained has been formalized into a variety of different psychotherapies. Among the best studied are *motivational enhancement therapy*, *cognitive behavioral therapy*, and *twelve-step facilitation therapy*. Each takes a unique approach to using the resources found in the human brain to counteract the destructive impulses of the malfunctioning desire-dopamine circuit.

MOTIVATIONAL ENHANCEMENT THERAPY: DESIRE DOPAMINE VERSUS DESIRE DOPAMINE

Addicts crave drugs. They use drugs even when drugs destroy their lives, but most of them know they're harming themselves. They're not completely deceived by the chemical. They're ambivalent: part of them wants nothing more than to use drugs, but there are other, weaker desires as well. Those desires can be strengthened. There may be a desire to be a better spouse, a better parent, or to do better at work. The drug addict may see their bank account drain away, and wish for the peace of mind that comes with financial security. Or they may wake up feeling sick every day, and wish they could go back to the time when they were strong and healthy.

None of these desires is able to provoke dopamine release the way drugs do, but desire not only gives us motivation to act; it also gives us patience to endure. In motivational enhancement therapy (MET), patients tolerate feeling resentful and deprived, the punishment of disappointed dopamine, because they know it will lead to something better. The goal of the therapy is to stoke the flames of desire for a better life.

MET therapists build up motivation by encouraging their patients to talk about their healthy desires. There's an old saying: "We don't believe what we hear, we believe what we say." For example, if you give someone a lecture on the importance of honesty, then have them play a game in which cheating is rewarded, you'll probably find that the lecture had little effect. On the other hand, if you ask someone to give *you* a lecture on the importance of honesty, they will be less likely to cheat when they sit down to play the game.

MET is a little manipulative. When the patient makes a statement the therapist likes, referred to as a *pro-change statement,* such as, "Sometimes I have trouble getting to work on time after a night of heavy drinking," the therapist responds with positive reinforcement, or a request to "tell me more about that." On the other hand, if the patient makes an *anti-change statement,* such as, "I work hard all day, and I deserve to relax in the evening with a few martinis," the therapist doesn't argue, because that would provoke more anti-change statements as the debate goes back and forth. Instead, she simply changes the subject. Patients usually don't notice what's going on, so the technique slips past their conscious defenses, and they spend the majority of the therapy hour making pro-change statements.

COGNITIVE BEHAVIORAL THERAPY: CONTROL DOPAMINE VERSUS DESIRE DOPAMINE

It's better to be smart than strong. Instead of trying to attack an addiction head on through willpower, cognitive behavioral therapy (CBT) uses the planning ability of control dopamine to defeat the raw power of desire dopamine. Addicts who struggle to stay clean are most often defeated when they are unable to resist craving. CBT therapists teach patients that craving is triggered by cues: drugs, alcohol, and things that remind the addict of drugs and alcohol (people, places, and things). Cues that suddenly and unexpectedly remind an addict of drugs produce a reward prediction error, like the addict who felt an overwhelming desire for heroin when he saw a bottle of laundry bleach. Desire dopamine cranks up, motivating the addict to use, and threatening to shut down completely if it doesn't get what it wants.

Alcoholics in CBT learn to arm themselves against cue-triggered craving in a number of different ways. For example, they may recruit a sober buddy to go with them to events where alcohol is being served. They also work to eliminate as many cues as possible. The patient and a friend are sent on a "search-and-destroy mission" in which everything that reminds the patient of alcohol is removed from his home: cocktail glasses, shakers, hip flasks, martini olives, and so forth. Anything that the drinker connects to alcohol use is a trigger, and has to go because otherwise it might be the agent of craving that brings an end to a hard-fought period of sobriety. One alcoholic patient brewed beer in his basement. He resisted getting rid of his beloved equipment, because it was his hobby, and had nothing to do with drinking, he said. Desire dopamine won that battle until he finally relented and threw everything in the garbage. Now he's sober.

ADDICTION: IT'S WORSE THAN YOU THINK

Addictions are hard to treat, harder than many other psychiatric illnesses. With other illnesses, such as depression, patients want to get better—there's no question about it. But if a person is addicted to a drug, he's not so sure. He may share the sentiment expressed by Saint Augustine while he was carrying on an affair with a young woman. He prayed, *Lord, give me chastity, but not yet.*

Because they're so difficult to overcome, doctors and patients often characterize addictive substances like alcohol as the enemy. It's an enemy we respect, because it's not only powerful, it's clever.

One "trick" is the use of unexpected triggers that lead to craving: photos taken with friends at a tailgate party, a favorite glass, a bottle opener, even a kitchen knife used to slice lemons. These triggers may be so subtle that the person may not recognize them until after they succumb to temptation.

But getting rid of the triggers isn't enough. Scientists have recently learned about a completely unexpected and somewhat frightening tactic the enemy has at its disposal. Consider an alcoholic who, for no apparent reason, decides to switch up his routine one day and take an alternate route home from work. He happens to pass a bar he used to go to and is overcome by craving. When he speaks about his relapse in his next therapy session, he has no idea how it happened. He doesn't connect the seemingly innocent decision to change his routine with the relapse.

But this relapse wasn't a coincidence. Scientists recently discovered that being addicted to alcohol changes the way certain segments of DNA work, segments that are essential for the normal functioning of the dopamine control circuits in the frontal lobes. A key enzyme is suppressed, interfering with the neurons' ability to transmit signals. It's like a hacker taking out the enemy's communication channels right in the middle of a battle. Thus an alcoholic may not want to drive past his old haunt, but the enemy has impaired his ability to appreciate the consequences of his decision to take the new route home.

The research that found the dangerous changes in DNA was done in rats, so we're not completely sure if the same thing happens in humans, but the results were striking. Rats with addiction-modified DNA drank more alcohol, and they drank even when the alcohol was spiked with quinine, which has a bitter taste that rats normally avoid. This finding suggested that the DNA alteration makes drinkers consume alcohol in spite of unpleasant consequences.

Alcoholics can still overcome their addiction, but impairing control dopamine's ability to oppose desire dopamine's impulses makes things difficult. Not only does alcohol create a perpetual desire; it also undermines the future-focus needed to stay on the road to recovery. The good news is we now know this weapon exists, and if we can find a way to reverse the DNA changes, we can neutralize it.

TWELVE-STEP FACILITATION THERAPY: H&N VERSUS DESIRE DOPAMINE

Alcoholics Anonymous (AA) is the most successful self-help fellowship in the world, but it's not for everyone. It requires people to accept the label of *alcoholic*, which many don't like. It's based on belief in a higher power, which some people don't have. And it requires sharing personal stories in a group setting, which makes some people uncomfortable. But those who fit in well can benefit from access to a valuable resource.

Overcoming addiction is a long-term battle, sometimes even lifelong. With that in mind, AA has some important advantages over drug treatment programs. AA has no limitations on how long a person can participate. AA is free and available all over the world, and in metropolitan areas there are groups all over the city that meet day and night.

AA is a fellowship rather than a treatment. A person gets better through relationships with other members of the group and their relationship with a higher power. The social part of our brain makes connections with other people using H&N neurotransmitters. There are few things in this world as powerful as relationships. According to Alexa, an internet analytic company, Facebook is the second most visited site on the web. (Google is number one, and Pornhub, the most visited pornography site, is all the way down at number 67, which should give us all faith in humanity's ability to resist the less healthy parts of desire dopamine.)

AA participants freely hand out their phone numbers so that struggling alcoholics have people to call for support and encouragement. If an AA member slips and experiences a relapse, no one condemns him, but he will inevitably feel like he has let them down. The H&N experience of guilt is a powerful motivator (as your mother knows). The combination of emotional support and the threat of guilt helps many addicts maintain a long-lasting sobriety.

A more dramatic example of H&N activity suppressing dopamine-driven addiction is the observation that when women smokers become pregnant, the quit rate spikes upward. Dr. Suena Massey of the Women's Health Research Institute at Northwestern University, who has done an in-depth study of this rapid change, notes that the usual steps a smoker goes through on the road to quitting are completely skipped. The level of H&N empathy for the developing fetus is so high that many women smokers jump right to the finish line and stop smoking without any conscious effort at all. Once the dopaminergic rationalization of *I'm not hurting anyone but myself* breaks down, the door opens for a rapid readjustment in the H&N–dopamine balance.

The dopamine system as a whole evolved to maximize future resources. In addition to desire and motivation, which get the ball rolling, we also possess a more sophisticated circuit that gives us the ability to think long term, make plans, and use abstract concepts such as math, reason, and logic. Looking into the longer-term future also gives us the tenacity we need to overcome challenges and accomplish things that take a long time, things like getting an education or flying to the moon. It also gives us the ability to tame the hedonistic urges of the desire circuit, suppressing immediate gratification to achieve something better. The control circuit suppresses H&N emotion, allowing us to think in a cold, rational way that's often required when hard decisions need to be made, such as sacrificing the well-being of one person for the benefit of others.

The control circuit can be crafty. Sometimes it charges straight ahead and dominates a situation through the power of confidence. Other times it leads to submissive behaviors that induce others to cooperate with us, multiplying our ability to get things done and reach our goals.

Dopamine yields not just desire but also domination. It gives us the ability to bend the environment and even other people to our will. But dopamine can do more than give us dominion over the world: it can create entirely new worlds, worlds that may be so astonishing, they could have been created only by a genius—or a madman.

FURTHER READING

MacDonald, G. (1993). *The light princess: And other fairy tales*. Whitethorn, CA: Johannesen.

Previc, F. H. (1999). Dopamine and the origins of human intelligence. *Brain and Cognition, 41*(3), 299–350.

Salamone, J. D., Correa, M., Farrar, A., & Mingote, S. M. (2007). Effort-related functions of nucleus accumbens dopamine and associated forebrain circuits. *Psychopharmacology, 191*(3), 461–482.

Rasmussen, N. (2008). *On speed: The many lives of amphetamine*. New York: NYU Press.

McBee, S. (1968, January 26). The end of the rainbow may be tragic: Scandal of the diet pills. *Life Magazine*, 22–29.

PsychonautRyan. (2013, March 9). Amphetamine-induced narcissism [Forum thread]. Bluelight.org. Retrieved from http://www.bluelight.org/vb/threads/689506-Amphetamine-Induced-Narcissism?s=e81c6e06edabb-cf704296e266b7245e4

Tiedens, L. Z., & Fragale, A. R. (2003). Power moves: Complementarity in dominant and submissive nonverbal behavior. *Journal of Personality and Social Psychology, 84*(3), 558–568.

Schlemmer, R. F., & Davis, J. M. (1981). Evidence for dopamine mediation of submissive gestures in the stumptail macaque monkey. *Pharmacology, Biochemistry, and Behavior, 14*, 95–102.

Laskas, J. M. (2014, December 21). Buzz Aldrin: The dark side of the moon. *GQ*. Retrieved from http://www.gq.com/story/buzz-aldrin

Cortese, S., Moreira-Maia, C. R., St. Fleur, D., Morcillo-Peñalver, C., Rohde, L. A., & Faraone, S. V. (2015). Association between ADHD and obesity: A systematic review and meta-analysis. *American Journal of Psychiatry, 173*(1), 34–43.

Goldschmidt, A. B., Hipwell, A. E., Stepp, S. D., McTigue, K. M., & Keenan, K. (2015). Weight gain, executive functioning, and eating behaviors among girls. *Pediatrics, 136*(4), e856–e863.

O'Neal, E. E., Plumert, J. M., McClure, L. A., & Schwebel, D. C. (2016). The role of body mass index in child pedestrian injury risk. *Accident Analysis & Prevention, 90*, 29–35.

Macur, J. (2014, March 1). End of the ride for Lance Armstrong. *The New York Times*. Retrieved from https://www.nytimes.com/2014/03/02/sports/cycling/end-of-the-ride-for-lance-armstrong.html

Schurr, A., & Ritov, I. (2016). Winning a competition predicts dishonest behavior. *Proceedings of the National Academy of Sciences, 113*(7), 1754–1759.

Trollope, A. (1874). *Phineas redux*. London: Chapman and Hall.

Power, M. (2014, January 29). The drug revolution that no one can stop. *Matter*. Retrieved from https://medium.com/matter/the-drug-revolution-that-no-one-can-stop-19f753fb15e0#.sr85czt5n

Baumeister, R. F., Bratslavsky, E., Muraven, M., & Tice, D. M. (1998). Ego depletion: Is the active self a limited resource? *Journal of Personality and Social Psychology, 74*(5), 1252–1265.

MacInnes, J. J., Dickerson, K. C., Chen, N. K., & Adcock, R. A. (2016). Cognitive neurostimulation: Learning to volitionally sustain ventral tegmental area activation. *Neuron, 89*(6), 1331–1342.

Miller, W. R. (1995). *Motivational enhancement therapy manual: A clinical research guide for therapists treating individuals with alcohol abuse and dependence*. Darby, PA: DIANE Publishing.

Kadden, R. (1995). *Cognitive-behavioral coping skills therapy manual: A clinical research guide for therapists treating individuals with alcohol abuse and dependence* (No. 94). Darby, PA: DIANE Publishing.

Nowinski, J., Baker, S., & Carroll, K. M. (1992). *Twelve step facilitation therapy manual: A clinical research guide for therapists treating individuals with alcohol abuse and dependence* (Project MATCH Monograph Series, Vol. 1). Rockville, MD: U.S. Dept. of Health and Human Services, Public Health Service, Alcohol, Drug Abuse, and Mental Health Administration, National Institute on Alcohol Abuse and Alcoholism.

Barbier, E., Tapocik, J. D., Juergens, N., Pitcairn, C., Borich, A., Schank, J. R., . . . Vendruscolo, L. F. (2015). DNA methylation in the medial prefrontal cortex regulates alcohol-induced behavior and plasticity. *The Journal of Neuroscience, 35*(15), 6153–6164.

Massey, S. (2016, July 22). An affective neuroscience model of prenatal health behavior change [Video]. Retrieved from https://youtu.be/tkng4mPh3PA

Creativity is the power to connect the seemingly unconnected.
—William Plomer, writer

Chapter 4

CREATIVITY AND MADNESS

The risks and rewards of the highly dopaminergic brain.

In which dopamine breaks down the barriers of the ordinary.

The same thoughts kept racing through my mind over and over and over again. I just wanted them to stop . . . Then I said, who am I going to call? Then I called up Ghostbusters. I mean, no, that came out wrong. I didn't call up Ghostbusters, I called crisis intervention . . . Can I go back inside now? I think somebody might be trying to shoot me.
—Excerpted from an interview with a man living with schizophrenia

The creative mind is the most potent force on earth. No oil well, gold mine, or thousand-acre farm can compete with the wealth-producing possibilities of a creative idea. Creativity is the brain at its best. Mental illness is the opposite. It reflects a brain struggling to manage even the most ordinary challenges of everyday life. Yet madness and genius, the worst and the best the brain can do, both depend on dopamine. Because of this basic chemical connection, madness and genius are more closely connected to each other than either is to the way ordinary brains work. Where does this connection come from, and what does it tell us about the essential nature of both? Let's start with madness.

BREAKING WITH REALITY

William had to be brought in by his parents because he refused to accept that he had a mental illness. His mother and father were both accomplished writers, and had traveled around the world visiting active war zones to collect material for their books. William had also shown signs of superior intelligence, although he was inconsistent. During his senior year of high school his parents promised to buy him a car if he got good grades, and he managed a 3.7 GPA.

Things changed dramatically after he went off to college. Strange ideas invaded his mind. He had made friends with a young woman, and he developed the mistaken belief that she was interested in him romantically. When she denied having these feelings, he came to the conclusion that she was HIV positive and was trying to protect him from infection. Soon, this idea spread to other people. He became convinced that more than a dozen people he knew were HIV positive, and that they were all counting on him to travel to Africa to find a cure. He figured this out because the voices of his dead grandmother and God were explaining things to him.

When his friends suggested he should see a mental health professional, William thought that his parents were bribing them to say this. It was part of a conspiracy, he thought, to make him think he was sick. He decided his parents were imposters, and he left the country to look for his real parents.

He didn't stay away long, but when he returned home, he accused his parents of monitoring him with hidden listening devices. He traveled to New York City to escape the overwhelming stress of his imagined persecution. He gave it the name "ambient abuse." Everything was becoming too intense, and he needed a break. He wanted to go someplace where no one could follow him.

By the time he returned home, paying a taxi driver $600 for the ride, his parents had had enough. They told him that he couldn't live in their house unless he saw a mental health specialist. William, who was now facing the prospect of becoming homeless, agreed. Under the supervision of a psychiatrist he began taking an antipsychotic medication. His condition improved, and he decided to enroll in a local community college, where he studied graphic design. It was early in his recovery, and the plan was too ambitious. After a few months he dropped out.

Over time, the medication progressively improved his symptoms, but it was a challenge for his parents to persuade him to take it on a regular basis. He continued to doubt that he had a psychiatric illness. His doctor switched William to a new drug that didn't require him to take pills every day. He just needed to come in once a month for an injection, allowing him to experience uninterrupted treatment. On this formulation he improved to the point where he was able to work full time as a cook and live independently in his own apartment.

Schizophrenia[1] is a form of psychosis notable for the presence of hallucinations and delusions. Hallucinations can cause a person to see things that aren't really there, feel their touch, even smell them. The most common type of hallucination is the auditory hallucination—hearing voices. The voices may comment on the person's behavior ("You're eating lunch now."). There may be more than one voice holding a conversation about the person ("Have you noticed that everybody hates him?" "It's because he doesn't shower."). Sometimes they're command hallucinations ("Kill yourself!"). Occasionally, the voices are friendly and encouraging ("You're a great guy. Keep up the good work.") Friendly hallucinations are the least likely to go away, which may be just as well. Overall, they have a positive influence.

---

1 "Madness" is not a psychiatric diagnosis. We use it here as it's used in conversation, meaning severe mental illness, including delusions and chaotic or disordered thoughts. The diagnosis most commonly referred to by the informal term *madness* is schizophrenia.

Another component of psychosis is delusions. These are fixed beliefs that are inconsistent with the generally accepted view of reality, such as "Aliens have implanted a computer chip in my brain." Delusions are held with absolute certainty, a level of certainty that is rarely experienced with nondelusional ideas. For example, most people are confident that their parents really are their parents, but if you ask them if they are absolutely certain, they will confess that they are not. On the other hand, when a schizophrenic patient was asked if he was sure that the FBI was using radio waves to implant messages in his head, he said there could be no doubt. No amount of evidence could convince him otherwise.

A good example of this phenomenon comes from John Nash, a Nobel Prize–winning mathematician, who lived with schizophrenia. Silvia Nasar, who wrote about Nash in her book *A Beautiful Mind*, recounted the following exchange between Nash and Harvard professor George Mackey:

> "How could you," began Mackey, "how could you, a mathematician, a man devoted to reason and logical proof . . . how could you believe that extraterrestrials are sending you messages? How could you believe that you are being recruited by aliens from outer space to save the world? How could you . . . ?"
>
> Nash looked up at last and fixed Mackey with an unblinking stare as cool and dispassionate as that of any bird or snake. "Because," Nash said slowly in his soft, reasonable southern drawl, as if talking to himself, "the ideas I had about supernatural beings came to me the same way that my mathematical ideas did."

Where, in fact, do these ideas come from? One clue comes from what we know about how to treat schizophrenia. Psychiatrists prescribe medications called antipsychotics that reduce activity within the dopamine desire circuit. At first glance, that seems odd. Stimulation of the desire circuit typically leads to excitement, wanting, enthusiasm, and motivation. How could excess stimulation cause psychosis? The answer comes from the concept of *salience*, a phenomenon that will also play a crucial role in understanding the roots of creativity.

SALIENCE AND THE DOPAMINE CONNECTION

Salience refers to the degree to which things are important, prominent, or conspicuous. One kind of salience is the quality of being unusual. For example, a clown walking down the street would be more salient—more out of place—than a man in a business suit. Another kind of salience is value. A briefcase with $10,000 in it is more salient than a wallet with $20. Different things are salient to different people. A jar of peanut butter is more salient to a boy with a peanut allergy than to one who is allergy free. It would also be more salient to a girl who loves peanut butter sandwiches compared to one who prefers tuna salad.

Think about how salient the following things are: a grocery store you've seen a hundred times before versus a grocery store that just opened yesterday, the face of a stranger versus the face of the person you secretly love, and a policeman as you are walking down the street versus a policeman after you just made an illegal left turn. Things are salient when they are important to you, if they have the potential to impact your well-being, for good or for evil. Things are salient if they have the potential to affect your future. Things are salient if they trigger desire dopamine. They broadcast the message, *Wake up. Pay attention. Get excited. This is important.* You're sitting at a bus stop, glancing at a newspaper article about a Canadian trade agreement. Unless the mind-numbing details of the negotiation will impact you in some way, your desire dopamine circuit is at rest. Then all of a sudden you come across the name of one of your classmates from high school. She's been involved in the negotiation of the pact. *Bang! Salience. Dopamine.* As you read further, your interest rising, you suddenly come across your own name. You can imagine how that would affect your dopamine.

A PSYCHOTIC SHORT CIRCUIT

What happens, though, if the salience function of the brain malfunctions—if it goes off even when there is nothing happening that is actually important to you? Imagine you're watching the news. The anchorman is talking about a government spying program, and suddenly your salience circuit fires for no reason at all. You might then believe that this story on the news has something to do with you. Too much salience, or any salience at all at the wrong time, can create delusions. The triggering event rises from obscurity to importance.

A common delusion among people with schizophrenia is that people on TV are talking directly to them. Another is that they are the target of investigation by the NSA, FBI, KGB, or Secret Service. One patient said he saw a stop sign, and thought it was a message from his mother telling him to stop looking at women. Another patient saw a red car parked outside her apartment on Valentine's Day, and believed it was a message of love from her psychiatrist. Even people who have never been psychotic might learn to attach salience to things that appear unimportant to others, such as black cats or the number 13.[2]

There's wide variation in how much salience different people attach to different things. Everyone has a lower limit, though. We have to categorize some things as having low salience, being unimportant, so we can ignore them for the simple reason that noticing every detail in the world around us would be overwhelming.

 BLOCKING DOPAMINE TO TREAT PSYCHOSIS

People with schizophrenia control their dopamine activity by taking medications that block dopamine receptors (Figure 4).

---

2   Is superstition a very mild form of delusion, or is it a choice? Research indicates that superstitious people have a preponderance of dopaminergic traits, so there is probably a genetic tendency for some people to adopt superstitious beliefs.

Receptors are molecules that sit on the outside of brain cells and catch neurotransmitter molecules (such as dopamine, serotonin, and endorphins). Brain cells have different receptors for different neurotransmitters, and each one affects the cell in a different way. Some receptors stimulate brain cells and others lull them into a state of tranquility. Changing cell behavior is how the brain processes information. It's similar to transistors turning on and off in a computer chip.

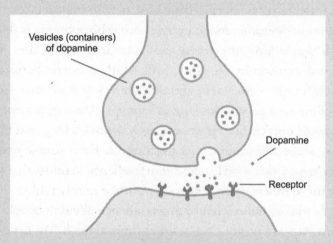

Vesicles (containers) of dopamine

Dopamine

Receptor

Figure 4

If something blocks a receptor, such as an antipsychotic medication, then the neurotransmitter (in this case, dopamine) can't get at it, and it can't communicate its signal. It's like putting a piece of tape over a keyhole. Blocking dopamine usually doesn't make all of the symptoms of schizophrenia go away, but it can get rid of the delusions and hallucinations. Unfortunately, antipsychotic medications block dopamine all over the brain, and blocking the control circuit in the frontal lobes can make certain aspects of the illness worse, such as difficulty paying attention and reasoning with abstract concepts.

Doctors try to maximize the benefits and minimize the harms by getting the dose just right. They want to suppress excess dopamine activity in the salience circuit without overly suppressing the control circuit, which is responsible for long-term planning. The goal is to give just enough medication to block 60 to 80 percent of the dopamine receptors. Additionally, when a dopamine surge occurs, signaling something important in the environment, it would be nice if the antipsychotic molecules got out of the way, just for a moment, to let the signal get through. If you're playing a video game, trying to defeat the boss, or applying for a new job, it would be nice to experience a little excitement to create the motivation that pushes things forward.

Older antipsychotic medications don't do this very well. They stick hard to the receptor. If something interesting happens and dopamine spikes, tough luck. The medication has latched on so tight, no dopamine can get through, and that doesn't feel good. Being cut off from natural dopamine surges makes the world a dull place and makes it hard to find reasons to get out of bed in the morning. Newer drugs bind more loosely. A surge of dopamine knocks the drug off the receptors, and the *this is interesting* feeling gets through.

DRINKING FROM A FIRE HOSE

In schizophrenia the brain short-circuits, attaching salience to ordinary things that ought to be familiar and therefore ignored. Another name for this is *low latent inhibition*. Typically, *latent* is used to describe things that are hidden, such as "a latent talent for music" or "a latent demand for flying cars." The way it's used in the phrase *latent inhibition* is somewhat different. It's not that a thing starts out hidden, it's that we *make* it hidden because it's not important to us.

We inhibit our ability to notice things that are unimportant so we don't have to waste our attention on them. If we're distracted by how clean the windows are as we walk down the street, we may miss the *Don't Walk* sign at the intersection. If we attach equal significance to the color of a person's tie and the expression on his face, we may fail to observe something very important to our future well-being. If you live next to a fire station, even the sounds of sirens will be inhibited after your dopamine circuits realize that nothing ever happens when they start to wail. Someone visiting your home might say, "What's that sound?" And you answer, "What sound?"

Sometimes our environment is so enriched with new things that latent inhibition is unable to pick and choose what is most important. This experience can be exhilarating or frightening depending on the situation and the person who is experiencing it. If you're in an exotic foreign country, there's not much to inhibit, and it can cause great pleasure but also confusion and disorientation—culture shock. Author and journalist Adam Hochschild described it this way: "When I'm in a country radically different from my own, I notice much more. It is as if I've taken a mind-altering drug that allows me to see things I would normally miss. I feel much more alive." As the new environment becomes familiar, we adjust, and eventually master it. We separate out the things that will affect us from those that won't, and latent inhibition returns, making us comfortable and confident in our new surroundings. We can once again separate the essential from the nonessential.

But what if the brain is unable to make this adjustment? What if the most familiar place feels like an alien environment? This problem is not confined to schizophrenia. A group of people living with this condition created a website called the Low Latent Inhibition Resource and Discovery Centre. They describe the feeling this way:

With low latent inhibition, an individual can treat familiar stimuli almost in the same manner as they would new stimuli. Think of the details you notice when you see something new for the first time and how it grabs your attention. From that all kinds of questions may arise in your mind. "What is that, what does it do, why is it there, what does it mean, how can it be utilised" and so on.

A visitor to the website described her experience in a comment:

> I'm losing my mind! There is just too much info in my head, and I get very little sleep. I can't stand to look at anything else! I'm tired of being an observer! I'm tired of seeing everything! . . . I want to go to the deep woods and see nothing, read nothing, drop all technology, watch nothing, hear nothing. I want no clutter, nothing moved, nothing changed. I want to sleep without dreams that give me answers to problems that put me back to work as soon as I get up! I'm tired and don't want to think anymore!

We see milder forms of low latent inhibition in the creative arts. Here's a simple example from the children's classic, *The House at Pooh Corner.* Winnie-the-Pooh, who is a poet, recites some verse to his small friend Piglet about Tigger, a boisterous new arrival to the Hundred Acre Wood. Piglet is a timid animal, and he points out how big Tigger is. Pooh thinks about what Piglet said, then adds a final stanza to his poem.

> *But whatever his weight in pounds,*
> *shillings, and ounces,*
> *He always seems bigger*
> *Because of his bounces.*

"And that's the whole poem," he said. "Do you like it, Piglet?"

"All except the shillings," said Piglet. "I don't think they ought to be there."

"They wanted to come in after the pounds," explained Pooh, "so I let them. It is the best way to write poetry, letting things come."

There may be chaos inside our heads that requires taming by the more logical parts of the brain, but there is also treasure. Whether or not you find that "shillings" improves Pooh's poem, one of the cardinal rules of creative writing is to turn off your inner censor when creating the first draft. If you're lucky, things will tumble out from your unconscious that will resonate in the unconscious of your readers, and your story will strike deep.

Here is a quotation from a schizophrenic patient that illustrates a more pathological tendency to "let things come."

> I got TV tooth, they call it. TV tooth is when they surprise you and put needles in your skull, and they listen to you for years if you know it or not. I didn't know it. They have this really fantastic, expensive equipment. They said to me, hey, we can check your head for, uh, if a bump shows up bruising, and the electricity is a little different across the top of your scalp, we'll guarantee social security for that injury or on its own. It's like cerebral palsy.

In this situation the speaker is unable to hold anything back. As thoughts come into his head, they are immediately translated into words with little processing. Normally, we pick and choose the things we say. We do this to censor unacceptable or illogical speech, but also to finish one thought before we begin the next. A close reading of the quotation makes it possible to get a general sense of what the speaker is saying, but it's hard.

With one thought rapidly taking the place of another, and a limited ability to hold the thoughts back, expression becomes highly disorganized. A less severe form of this type of jumping around is called *tangentiality*, in which the speaker leaps from one thought to another, but in a way that makes sense. For example, "I can't wait to go to Ocean City. They've got the best margaritas there. I have to find a place to get my car fixed this afternoon. Where are you going for lunch?" We often speak this way when we're excited. Desire dopamine gets revved up, and overwhelms control dopamine's more logical approach to communication.

At the far end of the spectrum is *word salad,* the most severe manifestation of out-of-control speech. In this case there is so much disorganization that there appears to be no sense to the utterance at all; for example, "How are you feeling this morning?" "Hospital pencils and ink newspaper critical care mother almost there."

They're selling postcards of the hanging
They're painting the passports brown
The beauty parlor is filled with sailors
The circus is in town
—"Desolation Row," Bob Dylan

Like people with mental illness, creative people such as artists, poets, scientists, and mathematicians will, at times, experience their thoughts running free. Creative thinking requires people to let go of the conventional interpretations of the world in order to see things in a brand-new way. In other words, they must break apart their preconceived models of reality. But what is a model, and why do we build them?

A WORLD BEYOND THE SENSES

Material things, objects in the H&N peripersonal space, can be experienced with all five senses. As an object moves away from us, from the peripersonal H&N to the extrapersonal dopamine, our ability to perceive it drops off one sensory modality at a time. First taste goes, then touch. As the thing moves farther away we lose our ability to smell it, hear it, and finally to see it. That's when things get interesting. How do we perceive something that is so far away that we can't even see it? We use our imagination.

Models are imaginary representations of the world that we build in order to better understand it. In some ways model building is like latent inhibition. Models contain only the elements of the environment that the model builder believes are essential. Other details are discarded. That makes the world easier to comprehend and, later, to imagine a variety of ways it might be manipulated for maximum benefit. Model building isn't something we're aware of. The brain builds models automatically as we go about our day, and updates them as we learn new things.

Models not only simplify our conception of the world; they also allow us to *abstract*, to take specific experiences and use them to craft broad, general rules. From this we can predict and deal with situations we've never encountered before. I may never have seen a Ferrari, but as soon as I do, I know it's for driving. I don't have to examine it, and run through all the different things I might do with it. It would be paralyzing if I had to do that with every car I encountered. Based on my experience with real cars, I built a model of an abstract car. If a car I've never seen before fits the general outlines of my abstract conception, I can quickly classify it and know that it's made for driving.

Recognizing a car may seem trivial, but model building also helps us with the most cosmic abstractions. Watching how real objects moved led Newton to develop his abstract law of universal gravitation, which not only predicts how apples fall from trees, but also the movements of planets, stars, and galaxies.

MENTAL TIME TRAVEL

Models can be helpful when we need to choose among a number of different options. They allow us to run through different scenarios in our imagination in order to select the best one. For instance, if I need to get from Washington, DC, to New York City I could take the train or the bus, or I could fly. To decide which will be fastest, most comfortable, or most convenient, I experience each option in my imagination, and then, based on my inner experience, I make my choice in the real world. This process is called *mental time travel*. Using imagination, we project ourselves into various possible futures, mentally experience them, then decide how we're going to get the most out of what we see—how we're going to maximize our resources, whether it's a roomy seat, a cheap ticket, or a fast ride.

Mental time travel is a powerful tool of the dopamine system. It allows us to experience a possible, though presently unreal, future as if we were there. Mental time travel depends on models because we make predictions regarding situations we haven't yet experienced. How would my life be different if I bought this new dishwasher? What sorts of problems might an astronaut face if he traveled to Mars? What would happen if I ran that red light?

Mental time travel is in constant use because it's the mechanism for making every conscious choice in life. To the brain, each deliberate choice about the future is a matter for the dopamine system and the models it has created, whether you are deciding what to order at Burger King or the president is deciding whether to start a war. Mental time travel is responsible for every "next step" in our lives.

HOW DID I END UP WITH SUCH A CRUMMY MODEL, AND CAN I FIX IT?

Before the psychiatrist met his patient, a twenty-year-old college student named May, her father came in to prepare the doctor for his first appointment with her. "She has never given us any trouble before," he said. "She's a good girl." May had been the perfect student. She was valedictorian of her high school class and had been admitted to a prestigious program of study at a nearby university. She had never gotten into trouble of any kind: no drugs, no alcohol, no staying out late. She had always been respectful toward her immigrant parents and had lived up to every expectation they had for her. Now she was being discharged from the hospital after a suicide attempt that had put her in the intensive care unit for a week.

When May came in for her initial visit, she was 30 minutes early and waited patiently in the reception area for her turn to see the doctor. She was slender, dressed as if she were going to a job interview. Her voice was hushed. Sometimes it was hard to hear what she was saying. It was as if she didn't believe the things she had to say were important enough to be spoken aloud.

May told the doctor that she couldn't concentrate, couldn't sleep, and sometimes cried for hours at a time. She had stopped going to class, and spent all day in her bedroom with the shades pulled down. It was clear that she couldn't function in the high-stress environment of the intensive course of study she had signed up for, and she had taken a leave of absence. More than anything else, she felt guilt. Always the perfect daughter, she now believed she was a source of shame to her family.

When May's family had first come to the United States she was just a girl, but she learned to speak English quickly, and became responsible for taking care of the entire family. She made sure the utility bills got paid. She called a plumber when the sink backed up. And when her parents fought, she was the referee. She believed that the happiness and success of her family lay on her shoulders. She had to be a straight-A student. She had to be thin and well-dressed. She wasn't allowed to rebel like other adolescents. She always had to do what she was told and was never allowed to disagree.

Her doctor expected her to respond well to treatment. She was cooperative and smart. But no matter what he did, nothing changed. Her depression wouldn't go away. When her leave of absence came to an end, May withdrew from school.

It was a long time before May revealed her secret. She was abusing amphetamines. It was the only way she could keep up with her studies, maintain a weight that was acceptable to her mother, and manage all the chores associated with the family responsibilities she had taken on. It worked for a while, but it was a coping mechanism that was destined to fail. There were emotional problems, too. Having missed out on normal teenage rebellion, a confusion of anger and resentment swirled inside her, and she didn't know what to do with those frightening feelings. Ultimately, the only possible treatment for her was moving to a different city. She needed to put many miles of distance between herself and her family before she could begin to figure out who she was.

How well our models fit the real world is of great importance. If our models are poor, we will make bad predictions about the future and subsequently bad choices. Poor models of reality may be caused by many things: not having enough information, difficulty with abstract thinking, or the stubborn persistence of wrong assumptions. Such bad assumptions may be so harmful that they lead to psychiatric illnesses such as anxiety and depression. For example, if a child grows up with critical parents, she may develop the conviction that she is an incompetent person, and this belief will shape the models of the world that she creates all her life. Therapists can address these faulty, often unconscious assumptions through psychotherapy, which may include insight-oriented psychotherapy, in which the patient and the therapist work to uncover suppressed memories that locked in the negative assumptions. Another helpful technique is CBT, which addresses the assumptions head on, and teaches the patient practical strategies for changing them.

As we gain experience with the world, we develop better and better models, and this is the basis of wisdom. We embrace models that work well, and discard the ones that fail to take us where we want to go. Knowledge passed on from previous generations can help us improve our models in a different way than direct experience. We have folk wisdom that tells us "a stitch in time saves nine," as well as the inherited knowledge of the great scientists and philosophers.

BREAKING MODELS: STARTING DOWN THE PATH OF CREATIVITY

If all you have is a hammer, everything looks like a nail.
—proverb

Models are powerful tools, but they have disadvantages. They can lock us in to a particular way of thinking, causing us to miss out on opportunities to improve our world. For instance, most people know that computers require instructions to work. Programmers type these instructions on a keyboard. This suggests a simple model: *typing instructions on a keyboard is the way to operate a computer.* The scientists at Xerox PARC had to free themselves from that model before they could invent the computer mouse and the graphical user interface. It's dopamine that builds models, and dopamine that breaks them apart. Both require us to think about things that don't currently exist, but might in the future.

Model breaking can be seen in certain kinds of riddles, called insight problems. Preexisting models have to be taken apart in order to see the problem in a fresh way. Here's an example:

I'm in years but not months. I'm in weeks but not days. What am I?

This riddle is difficult, and unless you've heard it before or have low latent inhibition, it's unlikely you'll figure out that the answer is the letter *e*. The riddle draws you into a calendar-based model, leading you to inhibit apparently irrelevant information, such as the letters that make up the words.

Here's another example. What one word does the sequence "HIJKLMNO" represent? A man who was puzzling over this problem experienced a series of dreams that were all about water. He wasn't able to make the connection, but it becomes obvious when we look at the answer: H_2O. We'll look more closely at the dopaminergic power of dreams later in the chapter.

Here's a riddle that a few decades ago required significant model breaking to find the solution. Today, it's much easier.

A father and his son are in a car accident. The father dies instantly, and the son is taken to the nearest hospital. The surgeon comes in and exclaims, "I can't operate on this boy. He's my son!" How is this possible?

DISCOVERING THE SOURCE OF CREATIVITY . . .

Oshin Vartanian, a researcher at York University in Toronto, wanted to figure out what part of the brain was most active when people discovered novel solutions to problems, so he scanned people's brains while they were solving a problem that required creativity. He found that when they discovered the solution to the problem, the front of their brains on the right side was activated. He wondered if this part of the brain was also involved in model breaking.

In a second experiment he asked participants not to solve a problem but simply to use their imagination. First he asked them to imagine real things, such as "a flower that is a rose." Then he asked them to imagine things that don't exist, things that don't fit the conventional model of reality, such as "a living thing that is a helicopter." With the volunteers in the scanner, he found that the same part of the brain lit up as before, but only when participants thought about objects that did not exist in life. When they imagined reality itself, the region stayed dark.

Brain scans of people with schizophrenia show changes in that same area, the right ventrolateral prefrontal cortex. Maybe it's because when we are being creative, we behave a little bit like a person with schizophrenia. We stop inhibiting aspects of reality that we had previously written off as unimportant, and we attach salience to things we once thought were irrelevant.

. . . AND SHOCKING IT TO LIFE

Finding the neural basis of creativity has enormous potential, because creativity is the most valuable resource in the world. New ways of growing crops feed millions of people. From candles to light bulbs, innovations in turning fuel into light have decreased its cost by a factor of a thousand. Might there be a way to boost this priceless treasure? Would it be possible for someone to become more creative if a scientist stimulated the parts of the brain that are active during creative thinking?

Researchers funded by the National Science Foundation decided to try. They used a technique called transcranial direct current stimulation (tDCS). As the name suggests, specific regions of the brain are stimulated using direct current (DC)—that's the kind of current you get from a battery, as opposed to alternating current (AC), which comes from a wall socket. DC is safer than AC and the amount of electricity used is small. Some devices are powered by a simple 9-volt battery, the boxy kind you put in your smoke detectors. tDCS machines can be very simple. Although commercial ones used for research cost over a thousand dollars, some brave individuals have cobbled together primitive ones using $15 worth of parts from their local electronics store. (Consumer tip: Don't do it.)

In small studies these devices have been shown to accelerate learning, enhance concentration, and even treat clinical depression. To attempt to enhance creativity, electrodes were attached to the foreheads of thirty-one volunteers, and the part of the brain that lies just behind the eyes was stimulated. Creativity was measured by testing the participants' ability to make analogies.

Analogies represent a very dopaminergic way of thinking about the world. Here's an example: light can sometimes act like individual bullets being fired from a gun, and at other times like ripples traveling across a pond. An analogy pulls out the abstract, unseen essence of a concept, and matches it with a similar essence of an apparently unrelated concept. The body's senses perceive two different things, but reason understands how they are the same. Pairing a brand-new idea with an old familiar one makes the new idea easier to understand.

The ability to draw a connection between two things that had previously appeared to be unrelated is an important part of creativity, and it appears that it can be enhanced by electrical stimulation. Compared to participants who were given fake tDCS, those who got electricity created more unusual analogies—that is, analogies between things that seemed very unlike one another. Nevertheless, these highly creative analogies were just as accurate as the more obvious ones created by the participants whose devices were secretly turned off.

Dopaminergic drugs can do the same thing. Although some patients who take dopaminergic drugs for Parkinson's disease develop devastating compulsions, others experience enhanced creativity. One patient who came from a family of poets had never done any creative writing. After starting dopamine-boosting drugs for his Parkinson's disease, he wrote a poem that won the annual contest of the International Association of Poets. Painters treated with Parkinson's medication often increase their use of vivid color. One patient who developed a new style after being treated said, "The new style is less precise but more vibrant. I have a need to express myself more. I just let myself go." Just like Winnie-the-Pooh: "It is the best way to write poetry, letting things come."

DREAMS: WHERE CREATIVITY AND MADNESS MINGLE

Few of us are geniuses or madmen, but we have all experienced the midpoint on this continuum: dreams. Dreams are similar to abstract thought in that they work with material taken from the external world, but they arrange the material in ways that are unconstrained by physical reality. Dreams often contain the theme of *up*, such as flying or falling from a great height. Dreams often involve future themes, too, sometimes in the form of the pursuit of some intensely desired goal that's always just out of reach. Abstract, detached from the real world of the senses, dreams are dopaminergic.

Freud named the mental activity that takes place in dreams "primary process," which is unorganized, illogical, created without regard to the limitations of reality, and driven by primitive desires. Primary process has also been used to describe the thought process seen in people with schizophrenia. As the German philosopher Arthur Schopenhauer wrote, "Dreams are brief madness and madness a long dream."

Dopamine is unleashed during dreaming, freed from the restraining influence of the reality-focused H&N neurotransmitters. Activity in the H&N circuits is suppressed because sensory input from the outside world into the brain is blocked. This freedom allows dopamine circuits to generate the bizarre connections that are the hallmark of dreams. The trivial, the unnoticed, and the odd can be elevated to positions of prominence, supplying us with new ideas that otherwise would have been impossible to discover.

The similarity between dreaming and psychosis has fascinated many researchers, and has spawned a rich scientific literature. A group from the University of Milan in Italy looked at the presence of bizarre thought content in the dreams of healthy people, and compared them to waking fantasies of both healthy participants and those with schizophrenia.

Scientists stimulated waking fantasies[3] using the Thematic Apperception Test (TAT), a series of cards showing ambiguous, sometimes emotionally charged pictures of people in various situations. Themes include success and failure, competition and jealousy, aggression, and sexuality. The participant is asked to study the picture, then make a story explaining the scene.

The Italian researchers compared the TAT stories and the descriptions of dreams of patients with schizophrenia to those of healthy comparison participants using a scale called the Bizarreness Density Index. The results of the tests confirmed that dreams are very much like psychosis. The Bizarreness Density Index was almost exactly the same for three categories of mental activity: (1) the descriptions of dreams of people with schizophrenia, (2) the waking TAT stories of people with schizophrenia, and (3) the descriptions of dreams of healthy people. On the other hand, the fourth category, waking TAT stories of healthy people, scored much lower on the index. This study supports Schopenhauer's conception that living with schizophrenia is like living in a dream.

---

3   In this context, *fantasy* refers broadly to the products of the imagination, rather than the more common use to signify daydreams of things like unlimited wealth.

HOW TO HARVEST CREATIVITY FROM A DREAM

If dreaming is similar to psychosis, how do we get back to our normal selves? Does it happen all at once, or does it take some time to restore logical thought patterns? If it takes time, are we a little bit insane while the transition occurs? Here's something else to consider: sometimes when we're asleep we dream, and other times we don't. As we make the transition from sleep to wakefulness, is our thought process different if we are waking from a dream or from dreamless sleep?

Researchers at New York University used the TAT to evaluate the kinds of stories people produced after they were woken from dreaming sleep and compared them to TAT stories produced after they were woken from non-dreaming sleep. They found that fantasies produced immediately after dreaming were more elaborate. They were longer, and contained more ideas. The imagery was more vivid, and the content was more bizarre. Here is an example of a story given by a healthy volunteer after being woken from a dreaming state. The volunteer was shown a picture of a boy looking at a violin:

> He's thinking over his violin. He makes a sad impression. Wait
> a minute! He's bleeding out of his mouth! And his eyes . . .
> seems like he's blind!

Another volunteer who had been woken from a dream was shown a picture of a young man, slouched on the floor, his head resting on a bench. There is a pistol on the floor next to him. Here is the response:

> There is a boy in a bed. He may be having some kind of
> emotional problem. He is nearly crying, or it may be he's
> laughing, maybe having a game. It could also be a girl.
> They're both dead. Or maybe it's a cat? There is something
> on the floor . . . keys, a flower, or maybe it's a toy, or a boat.

After being woken from a non-dreaming sleep, this same participant was shown another card, and wrote a notably less bizarre description, stating simply that it was "a boy wearing a shirt, who doesn't have any socks on. I don't see very much else."

Many people have had the experience of waking from a dream, feeling as if they were caught between two worlds. Thinking is more fluid, making leaps from topic to topic, unconstrained by the rules of logic. In fact, some people report that they experience their most creative thoughts in this crack between the two worlds. The H&N filter that focuses our attention on the external world of the senses has not yet been reengaged; dopamine circuits continue to fire unopposed, and ideas flow freely.

Friedrich August Kekulé became famous when he discovered the structure of the benzene molecule, an important industrial chemical of that time. Chemists had established that the molecule was composed of six carbon atoms and six hydrogen atoms, which came as a surprise. Usually molecules of this sort have more hydrogen atoms than carbon atoms. It was clear that whatever structure the molecule took, it wasn't an ordinary one.

The chemists tried to arrange the carbon atoms and the hydrogen atoms in all sorts of ways that wouldn't violate the rules of chemical bonding. They knew that carbon atoms could be strung together like beads on a string, and there could also be side branches coming off at right angles, but none of the structures they tried were consistent with the known properties of the benzene molecule. The nature of its true shape was a mystery. Kekulé described the moment of insight when he realized what that shape was:

"There I sat and wrote my [chemistry textbook], but it did not proceed well, my mind was elsewhere. I turned the chair to the fireplace and fell half asleep. Again the atoms gamboled before my eyes. Smaller groups this time kept modestly to the background. My mind's eyes, trained by visions of a similar kind, now distinguished larger formations of various shapes. Long rows, in many ways more densely joined; everything in movement, winding and turning like snakes. And look, what was that? One snake grabbed its own tail, and mockingly the shape whirled before my eyes. As if struck by lightning I awoke."

The vision of the snake with its tail in its mouth, the ancient ouroboros, led to the insight that the six carbon atoms of the benzene molecule formed a ring. Like the snake with its tail in its mouth—complete in and of itself—dreams are inner representations of inner ideas. Cut off from the senses, dreams allow dopamine to run free, unconstrained by the concrete facts of external reality.

Dr. Deirdre Barrett, a psychologist and dream researcher at Harvard Medical School, notes that it's not surprising that the answer to Kekulé's problem took a visual form. Much of the brain is every bit as active during dreaming as it is when it is awake, but there are crucial differences. Not surprisingly, the parts of the brain that filter seemingly irrelevant details, the frontal lobes, are shut down. But there is increased activity in an area called the secondary visual cortex. This part of the brain doesn't receive signals directly from the eyes, which receive no input during dreaming. Instead, it is responsible for processing visual stimuli. It helps the brain make sense of what the eyes are seeing.

Dreams are highly visual. In her book *The Committee of Sleep: How Artists, Scientists, and Athletes Use Dreams for Creative Problem Solving—and How You Can Too*, Dr. Barrett explains that just as Kekulé discovered the structure of benzene in a dreamlike state, ordinary people can use dreams to solve practical problems, too. Dr. Barrett put the problem-solving power of dreams to the test in a group of Harvard undergraduate students.

She asked them to choose a problem that was important to them. It could be personal, academic, or more general. Next she taught them dream incubation techniques. These are strategies people can use to increase the likelihood of having a problem-solving dream. The students wrote down their dreams every night for a week or until they believed they had solved their problem. The problems and the dreams were then submitted to a panel of judges who decided if the dream really did provide a solution.

The results were striking. About half the students had a dream related to their problem, and 70 percent of those who dreamed about the problem believed their dreams contained a solution. The independent judges mostly agreed. Among the students who had a dream about their problems, the judges deemed that about half offered a solution.

One of the students in the study was trying to decide what kind of career he would pursue after graduation. He had applied to two graduate programs in clinical psychology, both of which were in his home state of Massachusetts. He had also applied to two industrial psychology programs, one in Texas and the other in California. One night he dreamed he was in an airplane, flying over a map of the United States. The plane developed engine trouble, and the pilot announced that they needed to find a safe place to land. They were right over Massachusetts, and the student suggested that they land there, but the pilot said it was too dangerous to land anywhere in that state. When he woke up, the student realized that after spending his whole life in Massachusetts, it was time to move on. For him, the location of the graduate school was more important than the area of study. His dopamine circuits had provided him with a new perspective.

DREAMING STORIES AND SONGS

Dreaming is a frequent source of artistic creativity. Paul McCartney said he heard the melody for "Yesterday" in a dream. Keith Richards said he came up with the lyrics and riff for "Satisfaction" in a dream. "I dream colors, I dream shapes, and I dream sound," said Billy Joel in an interview with the *Hartford Courant* about his song "River of Dreams." "I woke up singing that one, and then it wouldn't go away." REM's Michael Stipe wrote lyrics for the band's breakthrough song, "It's the End of the World as We Know It (And I Feel Fine)," the same way. "I had had a dream about a party," he told *Interview* magazine. "Everyone at the party had names that started with the initials L. B. except for me. It was Lester Bangs, Lenny Bruce, Leonard Bernstein. That's how one verse of the song came about." Author Robert Louis Stevenson cited dreams as a source for *The Strange Case of Dr. Jekyll and Mr. Hyde,* and Stephen King says that his novel *Misery* came from a dream, too.

DREAM INCUBATION: HOW TO SOLVE PROBLEMS IN YOUR SLEEP

Choose a problem that's important to you, one that you have a strong desire to solve. The greater the desire, the more likely it is that the problem will show up in a dream. Think about the problem before you go to bed. If possible, put it in the form of a visual image. If it's a problem with a relationship, imagine the person it involves. If you're looking for inspiration, imagine a blank piece of paper. If you're struggling with some sort of project, imagine an object that represents the project. Hold the image in your mind, so it's the last thing you think of before you fall asleep.

Make sure you have a pen and paper next to your bed. As soon as you wake up from a dream, write it down, whether or not you think it's related to the problem. Dreams can be tricky, and the answer may be disguised. It's important to write down the dream immediately because the memory will evaporate in seconds if you begin to think about something else. Many people have had the experience of waking up from an intense dream, one that's overflowing with personal meaning, and then being unable to recall any of the details less than a minute later.

It may take a few nights before you find what you're looking for, and the solution you get from your dream may not be the *best* solution. But it will probably be a novel solution, one that approaches the problem from a new direction.

WHY NOBEL PRIZE WINNERS LIKE TO PAINT

The fine arts and the hard sciences have more in common than most people believe, because both are driven by dopamine. The poet composing lines about a hopeless lover is not so different from the phys-icist scribbling formulas about excited electrons. They both require the ability to look beyond the world of the senses into a deeper, more pro-found world of abstract ideas. Elite societies of scientists are filled with artistic souls. Members of the U.S. National Academy of Sciences are one and a half times more likely to have an artistic hobby compared to the rest of us. Members of the U.K. Royal Society are about twice as likely, and Nobel Prize winners are almost three times as likely. The better you are at managing the most complex, abstract ideas, the more likely you are to be an artist.

This similarity between art and science became especially impor-tant when a computer programming crisis occurred at the turn of the millennium. Computer programmers had developed the habit of abbreviating years by using only the last two digits (e.g., 99 for 1999) in order to conserve then-expensive memory space (and a few keystrokes). The programmers weren't thinking ahead to the next millennium, when 99 might mean 2099. Thousands of programs were at risk of crashing; not just web browsers and word processors, but also software that controlled airplanes, dams, and nuclear power plants. The Y2K problem, as it was known, affected so many systems that there weren't enough computer programmers to fix them all. By some reports, a few companies recruited out-of-work musicians because they were able to learn programming so easily.

WHY GENIUSES ARE JERKS

Music and math go together because elevated levels of dopamine often come as a package deal: if you are highly dopaminergic in one area, you're likely to be highly dopaminergic in others. Scientists are artists and musicians are mathematicians. But there's a downside. Sometimes having lots of dopamine is a liability.

High levels of dopamine suppress H&N functioning, so brilliant people are often poor at human relationships. We need H&N empathy to understand what's going on in other people's minds, an essential skill for social interaction. The scientist you meet at the cocktail party won't shut up about his research because he can't tell how bored you are. In a similar vein, Albert Einstein once said, "My passionate sense of social justice and social responsibility has always contrasted oddly with my pronounced lack of need for direct contact with other human beings." And "I love Humanity but I hate humans." The abstract concepts of social justice and humanity came easily, but the concrete experience of encountering another person was too hard.

Einstein's personal life reflected his difficulties with relationships. He was far more interested in science than people. Two years before he and his wife separated, he began an affair with his cousin, and eventually married her. Once again, he was unfaithful, cheating on his cousin with his secretary and possibly a half-dozen other girlfriends as well. His dopaminergic mind was both a blessing and a curse—the elevated levels of dopamine that allowed him to discover relativity was most likely the same dopamine that drove him from relationship to relationship, never allowing him to make the switch to H&N-focused, long-term companionate love.

Understanding how the brains of geniuses work provides further insight into the dopaminergic personality, and the different ways it can manifest itself. We've already seen the impulsive pleasure-seeker who has difficulty maintaining long-term relationships and is vulnerable to addiction. We've also seen the detached planner who would rather stay late at the office than enjoy time with friends. Now we see a third possibility: the creative genius—whether painter, poet, or physicist—who has so much trouble with human relationships that he may appear to be slightly autistic.[4] In addition, the dopaminergic genius is so focused on his internal world of ideas that he wears different-color socks, forgets to comb his hair, and generally neglects anything having to do with the real world of the here and now. Plato wrote about an incident in which Socrates, the ancient Greek philosopher, stood glued to one spot for an entire day and night, thinking about a problem, completely unaware of what was going on around him.

---

4   Autism is also associated with abnormally high levels of dopamine activity in the brain.

These three personality types appear to be very different on the surface, but they all have something in common. They're overly focused on maximizing future resources at the expense of appreciating the here and now. The pleasure seeker always wants more. No matter how much he gets, it's never enough. No matter how much he looks forward to some promised pleasure, he is incapable of finding satisfaction in it. As soon as it comes he turns his attention to what's next. The detached planner also has a future/present imbalance. Like the pleasure seeker he also has a constant need for more, but he takes a long-term view, chasing more abstract forms of gratification such as honor, wealth, and power. The genius lives in the world of the unknown, the not yet discovered, obsessed with making the future a better place through her work. Geniuses change the world—but their obsession often presents itself as indifference toward others.

 ## BENEVOLENT MISANTHROPES

Highly intelligent, highly successful, and highly creative people—typically, highly dopaminergic people—often express a strange sentiment: they are passionate about people but have little patience for them as individuals:

The more I love humanity in general the less I love man in particular. In my dreams, I often make plans for the service of humanity . . . Yet I am incapable of living in the same room with anyone for two days together . . . I become hostile to people the moment they come close to me.
—Fyodor Dostoyevsky

I am a misanthrope and yet utterly benevolent, have more than one screw loose yet am a super-idealist who digests philosophy more efficiently than food.
—Alfred Nobel

I love humanity but I hate people.
          —Edna St. Vincent Millay

Sometimes they even use nearly identical language:

I love mankind . . . it's people I can't stand.
—Charles Schulz (writing for Linus in *Peanuts*)

It may be unseemly but it is explainable. Highly dopaminergic people typically prefer abstract thinking to sensory experience. To them, the difference between loving humanity and loving your neighbor is the difference between loving the idea of a puppy and taking care of it.

THE TRAGIC CONSEQUENCES

There was almost certainly a genetic contribution to Einstein's dopaminergic traits. One of his two sons became an internationally recognized expert on hydraulic engineering. The other was diagnosed with schizophrenia at the age of twenty, and died in an asylum. Large population studies have also found a genetic component of a dopaminergic character. An Icelandic study that evaluated the genetic profile of over 86,000 people discovered that individuals who carried genes that placed them at greater risk for either schizophrenia or bipolar disorder were more likely to belong to a national society of actors, dancers, musicians, visual artists, or writers.

Isaac Newton, who discovered calculus and the law of universal gravitation, was one of those troubled geniuses. He had difficulty getting along with other people, and engaged in an infamous scientific quarrel with German mathematician and philosopher Gottfried Leibniz. He was secretive and paranoid and showed little emotion, to the point of ruthlessness. When he served as Master of the Royal Mint he caused many counterfeiters to be hanged despite the objections of his colleagues.

Newton was haunted by insanity. He spent hours trying to find hidden messages in the Bible, and wrote over a million words on religion and the occult. He pursued the medieval art of alchemy, obsessively searching for the philosopher's stone, a mythical substance that alchemists believed had magical properties and could even help humans achieve immortality. At the age of fifty, Newton became fully psychotic and spent a year in an insane asylum.

Based on the evidence, it seems likely that Newton had elevated levels of dopamine that contributed to his brilliance, his social problems, and his psychotic breakdown. And he's not alone. Many brilliant artists, scientists, and business leaders are thought or known to have had mental illness. They include Ludwig van Beethoven, Edvard Munch (who painted *The Scream*), Vincent van Gogh, Charles Darwin, Georgia O'Keeffe, Sylvia Plath, Nikola Tesla, Vaslav Nijinsky (the greatest male dancer of the early twentieth century, who once choreographed a ballet that started a riot), Anne Sexton, Virginia Woolf, chess master Bobby Fischer, and many others.

Dopamine gives us the power to create. It allows us to imagine the unreal and connect the seemingly unrelated. It allows us to build mental models of the world that transcend mere physical description, moving beyond sensory impressions to uncover the deeper meaning of what we experience. Then, like a child knocking over a tower of blocks, dopamine demolishes its own models so that we can start fresh and find new meaning in what was once familiar.

But that power comes at a cost. The hyperactive dopamine systems of creative geniuses put them at risk of mental illness. Sometimes the world of the unreal breaks through its natural bounds, creating paranoia, delusions, and the feverish excitement of manic behavior. In addition, heightened dopaminergic activity may overwhelm H&N systems, hampering one's ability to form human relationships and navigate the day-to-day world of reality.

For some, it doesn't matter. The joy of creation is the most intense joy they know, whether they are artists, scientists, prophets, or entrepreneurs. Whatever their calling, they never stop working. What they care about most is their passion for creation, discovery, or enlightenment. They never relax, never stop to enjoy the good things they have. Instead, they're obsessed with building a future that never arrives. Because when the future becomes the present, enjoying it requires activation of "touchy-feely" H&N chemicals, and that's something highly dopaminergic people dislike and avoid. They serve the public well. But no matter how rich, famous, or successful they become, they're almost never happy, certainly never satisfied. Evolutionary forces that promote the survival of the species produce these special people. Nature drives them to sacrifice their own happiness for the sake of bringing into the world new ideas and innovations that benefit the rest of us.

SURF, SAND, AND PSYCHOSIS

Brian Wilson of the Beach Boys is one of the most revolutionary popular musicians. In his early years, his music was deceptively simple: catchy tunes about surfing, cars, and girls. But as time went on, he conducted unprecedented sonic experiments—music just as pleasant to listen to, but successively more layered and complex. As a composer, arranger, and producer, he began to introduce new sounds and new combinations of sounds to pop music. Some of these choices were variations of familiar forms: unusual voicing of common chords, unlikely assemblies of tones as chords, and standard progressions that begin and end in unexpected places. Wilson employed unusual instruments such as the harpsichord and theremin, which was previously used to create the eerie humming noise in horror movies. He also used devices that were not considered musical instruments at all: a train whistle, bicycle bells, bleating goats. This experimentation culminated in the album *Pet Sounds* (1966), a critically acclaimed collection of creative music that sounded like nothing that had come before. If artists such as Bob Dylan elevated pop and rock lyrics from doggerel to poetry, Brian Wilson transformed the possibilities of the music itself from three chords and a verse-chorus structure to what Beach Boys publicist Derek Taylor is credited with calling the "pocket symphony."

The range of unusual creative connections suggests he experienced low latent inhibition associated with high levels of dopamine, but those high levels also may have contributed to Wilson's mental illness. "He hears voices," his wife Melinda Ledbetter told *People* magazine in 2012. "I can tell if it's good voices or bad voices by the look that comes over his face. For us it's hard to understand, but for him they're very real." He was diagnosed with schizophrenia, later changed to schizoaffective disorder, a combination of symptoms of schizophrenia and abnormal mood, including hallucinations and paranoia. In 2006, he told *Ability* magazine that he started hearing voices at the age of twenty-five, a week after he had taken psychedelic drugs. "For the past 40 years I've had auditory hallucinations in my head, all day every day, and I can't get them out. Every few minutes the voices say something derogatory to me . . . I believe they started picking on me because they are jealous. The voices in my head are jealous of me."

Wilson says that treatment to reduce the symptoms did not significantly reduce his creativity. Contrary to popular perception, the untreated pain of mental illness is a hindrance, not a help. "I used to go for long periods without being able to do anything, but now I play every day."

FURTHER READING

Orendain, S. (2011, December 28). In Philippine slums, capturing light in a bottle. *NPR All Things Considered.* Retrieved from https://www.npr.org/2011/12/28/144385288/in-philippine-slums-capturing-light-in-a-bottle

Nasar, S. (1998). *A beautiful mind.* New York, NY: Simon & Schuster.

Dement, W. C. (1972). *Some must watch while some just sleep.* New York: Freeman.

Winerman, L. (2005). Researchers are searching for the seat of creativity and problem-solving ability in the brain. *Monitor on Psychology, 36*(10), 34.

Green, A. E., Spiegel, K. A., Giangrande, E. J., Weinberger, A. B., Gallagher, N. M., & Turkeltaub, P. E. (2016). Thinking cap plus thinking zap: tDCS of frontopolar cortex improves creative analogical reasoning and facilitates conscious augmentation of state creativity in verb generation. *Cerebral Cortex, 27*(4), 2628–2639.

Schrag, A., & Trimble, M. (2001). Poetic talent unmasked by treatment of Parkinson's disease. *Movement Disorders, 16*(6), 1175–1176.

Pinker, S. (2002). Art movements. *Canadian Medical Association Journal, 166*(2), 224.

Gottesmann, C. (2002). The neurochemistry of waking and sleeping mental activity: The disinhibition-dopamine hypothesis. *Psychiatry and Clinical Neurosciences, 56*(4), 345–354.

Scarone, S., Manzone, M. L., Gambini, O., Kantzas, I., Limosani, I., D'Agostino, A., & Hobson, J. A. (2008). The dream as a model for psychosis: An experimental approach using bizarreness as a cognitive marker. *Schizophrenia Bulletin, 34*(3), 515–522.

Fiss, H., Klein, G. S., & Bokert, E. (1966). Waking fantasies following interruption of two types of sleep. *Archives of General Psychiatry, 14*(5), 543–551.

Rothenberg, A. (1995). Creative cognitive processes in Kekulé's discovery of the structure of the benzene molecule. *American Journal of Psychology, 108*(3), 419–438.

Barrett, D. (1993). The "committee of sleep": A study of dream incubation for problem solving. *Dreaming, 3*(2), 115–122.

Root-Bernstein, R., Allen, L., Beach, L., Bhadula, R., Fast, J., Hosey, C., & Podufaly, A. (2008). Arts foster scientific success: Avocations of Nobel, National

Academy, Royal Society, and Sigma Xi members. *Journal of Psychology of Science and Technology*, *1*(2), 51–63.

Friedman, T. (Producer), & Jones, P. (Director). (1996). *NOVA: Einstein Revealed*. Boston, MA: WGBH.

Kuepper, H. (2017). Short life history: Hans Albert Einstein. Retrieved from http://www.einstein-website.de/biographies/einsteinhansalbert_content. html

James, I. (2003). Singular scientists. *Journal of the Royal Society of Medicine*, *96*(1), 36–39.

Conservative: A statesman who is enamored of existing evils, as distinguished from the liberal, who wishes to replace them with others.
—Ambrose Bierce, *The Devil's Dictionary*

Chapter 5
POLITICS

Why we can't just get along.

In which we learn how superpowers and hand sanitizers affect our political ideology

THE AUTHORS REGRET . . .

In April 2002 the *American Journal of Political Science* published a research report, "Correlation not Causation: The Relationship Between Personality Traits and Political Ideologies." It was written by a group of researchers from Virginia Commonwealth University who studied the link between political beliefs and personality traits. They found that the two were connected, and that the connection could be attributed to genes. Along the way, they noticed that certain personality traits were associated with liberals and others with conservatives.

They were particularly interested in a collection of personality features—what psychiatrists call a *personality constellation*—called P. The authors noted that people with low *P* scores are more likely to be "altruistic, well socialized, empathic, and conventional." By contrast, people who have high *P* scores are "manipulative, tough-minded, and practical," and present characteristics such as "risk-taking, sensation-seeking, impulsivity, and authoritarianism." They concluded, "As such, we expect higher *P* scores to be related to a more conservative political attitude."

What they predicted was exactly what they found. The stereotypes, they said, were true: conservatives tend to be impulsive and authoritarian while liberals tend to be well socialized and generous. But in science, when you find just what you expect, it can be a red flag. And in January 2016, fourteen years after the original report, the journal published a retraction:

> The authors regret that there is an error in the published
> version of "Correlation not Causation: The Relationship
> Between Personality Traits and Political Ideologies." The
> interpretation of the coding . . . was exactly reversed.

Somebody had flipped the labels. The correct interpretation was the opposite of what they reported. It was the liberals in their study—not the conservatives—who were manipulative, tough-minded, and practical. And it was the conservatives, not the liberals, who tended to be altruistic, well socialized, empathic, and conventional. Many people expressed surprise at this reversal. But if we consider what the study found at its most basic level and how it relates to the dopamine system, the revised results make good sense—certainly more sense than the original findings, which were widely heralded but exactly backward.

THE LIMITATIONS OF PERSONALITY MEASURES

Psychologists have worked for decades to develop ways to measure personality. They found that personality can be divided into different domains, such as how open a person is to new experiences or how self-disciplined he is. American psychologists divide personality into five domains, while the British prefer three. Either way, when a scientist focuses on one of the domains, she is measuring only a slice of a person's personality, not the whole person. Consider two nurses who both have high compassion scores. At first glance, one might imagine two similar people. But there are other personality domains as well. One nurse might be outgoing and emotional, while the other is introverted and restrained. Even though nurses may have some personality features in common, they are a group made up of unique individuals.

Another limitation of personality measurements is that scientists usually report a group's *average* score. So if a study finds that liberals are more risk-taking than conservatives, it's likely that within that group of liberals there are some who are safety-seeking. Studies of personality help us predict what a group of people will do, but they are less helpful in predicting what an individual will do.

PROGRESSIVES IMAGINE A BETTER FUTURE

The characteristics the study eventually associated with liberals—risk-taking, sensation-seeking, impulsivity, and authoritarianism—are the characteristics of elevated dopamine.[1] But do dopaminergic people

---

1 In fact, a group of scientists from the Institute of Psychiatry in London found that dopamine receptors were crowded together more tightly in the brains of people with high *P* scores compared to those with lower scores. Dense receptor packing led to stronger dopamine signals, which in turn led to the emergence of the distinctive personality features. The connection is also seen when we look at what *P* stands for: *Psychoticism.* High *P* scores are a risk factor for the development of schizophrenia. That doesn't mean that all liberals are at risk for becoming psychotic, but many of them share things in common with highly creative people, who sometimes do tip into the realm of psychosis.

really tend to support liberal policies? It seems that the answer is yes. Liberals often refer to themselves as *progressives*, a term that implies constant improvement. Progressives embrace change. They imagine a better future and in some cases even believe that the right combination of technology and public policy can eliminate fundamental problems of the human condition such as poverty, ignorance, and war. Progressives are idealists who use dopamine to imagine a world far better than the one we live in today. Progressivism is an arrow pointing forward.

The word *conservative*, on the other hand, implies maintaining the best of what we have inherited from those who came before us. Conservatives are often suspicious of change. They don't like experts who try to advance civilization by telling them what to do, even when it's in their own best interest; for example, laws that require motorcyclists to wear helmets, or regulations that promote healthy eating. Conservatives distrust the idealism of progressives, criticizing it as an impossible effort to build a perfect utopia; an effort that is more likely to lead to totalitarianism in which the elite dominate all aspects of public and private life. In contrast to the arrow of progressivism, conservatism is better represented by a circle.

Matt Bai, former chief political correspondent for the *New York Times Magazine*, inadvertently recognized the dopamine difference between left and right when he wrote, "Democrats win when they embody modernization. Liberalism triumphs only when it represents a reforming of government, rather than the mere preservation of it . . . Americans don't need Democrats to stand up for nostalgia and restoration. They already have Republicans for that."

The connection between dopamine and liberalism is further demonstrated by looking at specific groups of people. Dopaminergic people tend to be creative. They also work well with abstract concepts. They like to pursue novelty and have a general dissatisfaction with the status quo. Is there any evidence that this type of person is likely to be politically liberal? Silicon Valley start-up companies attract exactly this type of person: creative, idealistic, skilled in abstract fields such as engineering, mathematics, and design. They are rebels, driven to pursue change, even at the risk of going broke. Silicon Valley entrepreneurs, and the people who work for them, tend to be quite dopaminergic. They are tough-minded, risk-taking, sensation-seeking, and practical—personality features associated with liberals in the corrected version of the *American Journal of Political Science* article.

What do we know about the politics of Silicon Valley? A survey of startup founders revealed that 83 percent held the progressive view that education can solve all or most of the problems in society. Among the general public, only 44 percent believe this is true. Startup founders were more likely than the general public to believe that government should encourage smart personal decisions. Eighty percent of them believed that almost all change is good over the long run. And in the 2012 presidential election, over 80 percent of employee donations from top tech firms went to Barack Obama.

FROM HOLLYWOOD TO HARVARD

Another example of the link between dopamine and liberalism can be found in the entertainment industry. Hollywood is the mecca of American creativity, as well as the model of dopaminergic excess. Our highest-profile celebrities feverishly pursue *more*: more money, more drugs, more sex, and whatever happens to be the latest fashion. They're easily bored. According to a study done by the Marriage Foundation, a U.K. think tank, the divorce rate among celebrities is almost twice that of the general population. It's even worse during the first year of marriage when couples must make the transition from passionate to companionate love. Newly married celebrities are almost six times as likely to divorce compared to ordinary people.

Many of the problems actors face are dopaminergic in nature. A 2016 study of Australian actors found that despite "feelings of personal growth and a sense of purpose in the actors' work," they were highly vulnerable to mental illness. The actors identified a number of key issues including "problems with autonomy, lack of environmental mastery, complex interpersonal relationships and high self-criticism." These are challenges that would be most difficult for highly dopaminergic individuals, who need to feel in control of their environment and often have difficulty navigating complex human relationships.

As for politics, liberal views dominate Hollywood. According to CNN, celebrities donated $800,000 to President Barack Obama's reelection campaign, compared to just $76,000 to Republican challenger Mitt Romney. The Center for Responsive Politics, which publishes the website Opensecrets.org, reported that during this same election cycle, people who worked for the seven major media corporations donated six times as much to Democrats as they did to Republicans.

Next on the list is academia. Academia is a temple of dopamine. Academics are described as living in an *ivory tower* (as opposed to an earthen hut, for example). They devote their lives to the immaterial, abstract world of ideas. And they are very liberal. You're more likely to find a communist than a conservative in academia. A *New York Times* opinion piece noted that only 2 percent of English professors were Republicans, while 18 percent of social scientists identified themselves as Marxist.

The enforcement of liberal orthodoxy is more widespread on university campuses than in any other setting. Comedian Chris Rock told a reporter for *The Atlantic* that he won't perform on college campuses because the audience is too easily offended by speech that runs counter to liberal ideology. Jerry Seinfeld also said in a radio interview that other comedians had told him not to go near colleges. "They're so PC," he was warned.

ARE LIBERALS SMARTER?

A career in academia is generally a sign of superior intelligence, but does superior intelligence extend to liberals in general, to people more likely to have highly active dopamine systems? It probably does. Testing the ability to manipulate abstract ideas, courtesy of the dopamine control circuit, is a fundamental part of how psychologists measure intelligence.

To explore the question of the relative intelligence of liberals and conservatives, Satoshi Kanazawa, a scientist at the London School of

Economics and Political Science, evaluated a group of men and women who had taken IQ tests back when they were in high school. The scores were averaged by political ideology, and a remarkably clear trend emerged. Adults who described themselves as *very liberal* had higher intelligence scores compared to those who described themselves as simply *liberal*. The liberals had higher scores than those who described themselves as *middle of the road*, and the progression held steady all the way down to those who described themselves as *very conservative*. With a score of 100 representing the average, very liberal adults had an IQ of 106 and very conservative ones had an IQ of 95.

A smaller but similar trend was seen with regard to religiosity. Atheists had an IQ of 103, whereas those who described themselves as very religious averaged 97. It's important to emphasize that these are averages. Within the larger groups there are brilliant conservatives and not-so-brilliant liberals. Furthermore, the overall differences are small. The "Normal" range is 90 to 109. "Superior intelligence" starts at 110 and "Genius" at 140.

Mental flexibility—the ability to change one's behavior in response to changing circumstances—is also an ingredient in how we measure intelligence. To evaluate mental flexibility, researchers at New York University set up an experiment in which they asked test participants to press a button when they saw the letter W and to refrain from pressing when they saw the letter M. The participants had to think fast. When the letter was displayed, they had only half a second to decide whether or not to press the button. To make things even harder, the researchers sometimes switched up the rule: press on M, refrain on W.

Conservatives had more difficulty than liberals, particularly when a series of press signals was followed by a refrain signal. When the signal for change came, they had trouble adjusting their behavior.

To get a better understanding of what was going on, the scientists attached electrodes to the participants' heads so they could measure brain activity during the test. There wasn't much difference between liberals and conservatives when the press symbol was displayed. But when the no-press signal came up, and the participants had half a second to make a decision, the liberals instantly fired up a part of their brain responsible for error detection (involving anticipation, attention, and motivation) in a way the conservatives did not. When circumstances change, liberals do a better job of rapidly activating neural circuits and adjusting their responses to meet the new challenge.

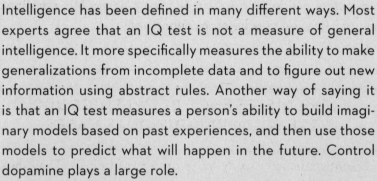

WHAT IS INTELLIGENCE?

Intelligence has been defined in many different ways. Most experts agree that an IQ test is not a measure of general intelligence. It more specifically measures the ability to make generalizations from incomplete data and to figure out new information using abstract rules. Another way of saying it is that an IQ test measures a person's ability to build imaginary models based on past experiences, and then use those models to predict what will happen in the future. Control dopamine plays a large role.

However, there are other ways to define intelligence, such as the ability to make good day-to-day decisions. For this type of mental activity emotions (H&N) are essential. Antonio Damasio, a neuroscientist at the University of Southern California and the author of *Descartes' Error: Emotion, Reason, and the Human Brain*, notes that most decisions cannot be approached in a purely rational way. Either we don't have enough information or we have far more than we can process. For example: Which college should I attend? What's the best way to tell her I'm sorry? Should I be friends with this person? What color should I paint the kitchen? Should I marry him? Is now a good time to express my opinion, or would it be better to keep quiet?

Being in touch with our emotions and processing emotional information skillfully are crucial for almost every decision we make. Intellectual prowess is not enough. Everyone is familiar with the scientific genius or brilliant writer who is like a helpless child in real life because he lacks "common sense"—the ability to make good decisions.

The role of emotions in decision making has not been studied as extensively as the role of rational thought; however, it wouldn't be unreasonable to predict that individuals who have a strong H&N system would have an advantage in this area. A high score on an IQ test may be a good predictor of academic success, but for a happy life, emotional sophistication may be more important.

THE DIFFERENCE BETWEEN GROUP TRENDS AND INDIVIDUAL CASES

Scientists usually study large groups of people. They measure characteristics they are interested in and calculate average values. Then they compare those averages with what's called a control group. A control group might be ordinary people, healthy people, or the general population. For example, a scientist might do a population study that reveals a higher rate of cancer among people who smoke cigarettes compared to everybody else. She might also do a genetics study and find out that people who have a gene that revs up the dopamine system are on average more creative compared to people who don't have that gene.

The problem is that when we talk about the averages of a large group, there are always exceptions, sometimes lots of exceptions. Many of us can think of heavy smokers who lived well into their nineties. Similarly, not everyone with a highly dopaminergic gene is creative.

Many things influence human behaviors: how dozens of different genes interact with one another, what kind of family you grew up in, and whether you were encouraged to be creative at a young age, to name a few. Having one specific gene usually has only a small effect. So while these studies advance our understanding of how the brain works, they're not very good at predicting how a particular individual—one member of that large group—will behave. In other words, some observations about a group you belong to may not be true about you in particular. That's to be expected.

RECEPTOR GENES AND THE LIBERAL-CONSERVATIVE SPLIT

There's a good chance that the difficulty the conservatives faced stemmed from differences in their DNA. In fact, political attitudes in general seem to be influenced by genetics. In addition to the *American Journal of Political Science* article just discussed, other studies support a link between a genetic disposition to a dopaminergic personality and a liberal ideology. Researchers from the University of California, San Diego focused on a gene that codes for one of the dopamine receptors called D4. Like most genes, the D4 gene has a number of variants. Slight variations in genes are called *alleles*. Each person's collection of different alleles (along with the environment they grew up in) helps determine their unique personality.

One of the variants of the D4 gene is called 7R. People who have the 7R variant tend to be novelty-seeking. They have less tolerance for monotony and pursue whatever is new or unusual. They can be impulsive, exploratory, fickle, excitable, quick-tempered, and extravagant. On the other hand, people with low novelty-seeking personalities are more likely to be reflective, rigid, loyal, stoic, slow-tempered, and frugal.

The researchers found a connection between the 7R allele and adherence to liberal ideology, but only if a person grew up around people with a variety of political opinions. There had to be both a genetic piece and a social piece for the connection to take place. A similar association was found among a sample of Han Chinese university students in Singapore, indicating that the link between the 7R allele and adherence to liberal ideology is not unique to Western culture.

HUMANS OR HUMANITY?

While conservatives on average may lack some of the virtuoso talents of the dopaminergic left, they are more likely to enjoy the advantages of a strong H&N system. These include empathy and altruism—particularly in the form of charitable giving—and the ability to establish long-term, monogamous relationships.

The left–right disparity in charitable giving was described in a research report published by *The Chronicle of Philanthropy*. The researchers used IRS data to evaluate charitable giving by state based on how each one voted in the 2012 election.[2]

The *Chronicle* found that people who gave the highest percentage of their incomes lived in states that voted for Romney, while people who gave the lowest percentage of their incomes lived in states that voted for Obama. In fact, every one of the top sixteen states for giving as a percentage of income voted for Romney. A breakdown by city found that the liberal cities of San Francisco and Boston were near the bottom, whereas Salt Lake City, Birmingham, Memphis, Nashville, and Atlanta were among the most generous. The differences were independent of income. Poor, rich, and middle-class conservatives all gave more than their liberal counterparts.

---

2   There were some weaknesses in the data. Since it came from tax returns, it relied on the 35 percent of taxpayers who itemize, and typically, it's wealthier taxpayers who itemize. Additionally, only about a third of charitable contributions go to the poor. According to a 2011 report from Giving USA, 32 percent of donations went to religious organizations, and 29 percent went to educational institutions, private foundations, arts, culture, and environmental charities. In spite of these weaknesses, the report provided an interesting overview of who is most likely to give money to others.

These results don't mean that conservatives care about the poor more than liberals do. Instead, it may be that, like Albert Einstein, liberals are more comfortable focusing on humanity rather than humans. Liberals advocate for laws that provide assistance to the poor. Compared to charitable giving, legislation is a more hands-off approach to the problem of poverty. This reflects our often-observed difference in focus: dopaminergic people are more interested in action at a distance and planning, while people with high H&N levels tend to focus on things close at hand. In this case, the government acts as the agent of liberal compassion and also serves as a buffer between the benefactor and the beneficiary. Resources for the poor are provided by bureaucracies that are funded collectively by millions of individual taxpayers.

Which is better: policy or charity? It depends on how you look at it. As one would expect, the dopaminergic approach, policy, maximizes resources that are made available to the poor. Maximizing resources is what dopamine does best. In 2012, federal, state, and local governments spent about $1 trillion on antipoverty programs. That's approximately $20,000 for every poor person in America. Charitable giving, on the other hand, was only $360 billion. The dopaminergic approach provided almost three times as much money.

On the other hand, the value of help is more than dollars and cents. The here-and-now emotional impact of impersonal government assistance is different from a personal connection with a church or charity. Charity is more flexible than law, so it's better able to focus on the unique needs of real individuals as opposed to abstractly defined groups. People who work for private charities typically come in close contact with the people they help, often actual physical contact. This intimate relationship allows them to get to know the people they help, and individualize the assistance that's provided. In this way, material resources can be augmented with emotional support, such as helping the able-bodied move toward employment or, more generally, showing the underserved that another person really does care about them as individuals. Many charities stress personal responsibility and good character as the most effective combatants of poverty. This approach will not work for everyone, but for some it will be more helpful than receiving government entitlements.

There is also an emotional benefit for the giver. The *hedonistic paradox* states that people who seek happiness for themselves will not find it, but people who help others will. Altruism has been associated with greater well-being, health, and longevity. There is even evidence that helping others slows down aging at the cellular level. Researchers in the Department of Bioethics at Case Western Reserve University suggest that the benefits of altruism may derive from "deeper and more positive social integration, distraction from personal problems and the anxiety of self-preoccupation, enhanced meaning and purpose in life, and a more active lifestyle." These are benefits that can't be achieved by merely paying taxes.

If policy directs more resources to the poor, and charity adds additional benefits, why not just do both? The problem is that dopamine and H&N neurotransmitters generally oppose each other, which creates an either/or problem. People who support government assistance for the poor (a dopaminergic approach) are less likely to give (an H&N approach) and vice versa.

The University of Chicago's General Social Survey has been tracking trends, attitudes, and behaviors in American society since 1972. One section of the survey asks questions about income inequality. The results showed that Americans who strongly oppose redistribution by government to address this problem gave 10 times more to charity than those who strongly support government action: $1,627 annually versus $140. Similarly, compared to people who want more welfare spending, those who believe that the government spends too much money on welfare are more likely to give directions to someone on the street, return extra change to a cashier, and give food or money to a homeless person. Almost everyone wants to help the poor. But depending on whether they have a dopaminergic or H&N personality, they will go about it in different ways. Dopaminergic people want the poor to receive *more* help, while H&N people want to provide personal help on a one-to-one basis.

COUPLING CONSERVATIVES

The preference for close, personal contact that leads conservatives to take a more hands-on approach to helping the poor also makes them more likely to establish long-term, monogamous relationships. The *New York Times* reported that "spending childhood nearly anywhere in blue America—especially liberal bastions like New York, San Francisco, Chicago, Boston and Washington—makes people about 10 percentage points less likely to marry relative to the rest of the country." In addition, when liberals marry, they're more likely to cheat.

In addition to charitable giving, the General Social Survey also tracks the sexual behavior of Americans. Starting in 1991 they began asking the question, "Have you ever had sex with someone other than your husband or wife while you were married?" Dr. Kanazawa, who evaluated the relationship between political ideology and intelligence, analyzed this data to find out who was most likely to answer *yes* to the question. Among those who identified as conservative, 14 percent had cheated on their spouses. The number fell slightly to 13 percent among those who identified as very conservative. Among liberals 24 percent reported cheating, and 26 percent of those who described themselves as very liberal reported cheating. The same trend was seen when the data for men and women were analyzed separately.

Conservatives have less sex than liberals, possibly because conservatives are more likely to be in companionate relationships in which testosterone is suppressed by oxytocin and vasopressin. Though the sex may be less frequent, it's more likely to end in orgasm for both partners. According to a study called "Singles in America," a survey of more than 5,000 adults designed by the University of Binghamton's Institute for Evolutionary Studies, conservatives are more likely to experience climax during sex than liberals.

Dr. Helen Fisher, chief scientific advisor at Match.com, speculated that conservatives are better at giving up control, an activity necessary for orgasm to occur. She attributed this ability to having clearer values, which makes it easier to relax. This explanation, which relies on a connection between clear values and disinhibition during climax, may not be the most straightforward explanation. There may be simpler ones based on what we know about the neurobiology of sex. Most obviously, letting go of control, which is necessary for climax to occur, is easier within a trusting relationship. This type of relationship is more common among stability-seeking H&N conservatives compared to novelty-seeking dopaminergic liberals. Additionally, the ability to enjoy the physical sensations of sex in the here and now requires suppression of dopamine by H&N neurotransmitters such as endorphins and endocannabinoids. Greater activity in the H&N system relative to dopamine makes that shift easier to achieve.

The dating website OkCupid did their own survey on sex, and found an intriguing piece of data with regard to what kind of people valued, or did not value, orgasms. They asked, "Are orgasms the most important part of sex?" They divided up the data based on political and professional affiliation. Those most likely to answer *no* to the question were politically liberal writers, artists, and musicians.

If you're highly dopaminergic—as writers, artists, and musicians tend to be—the most important part of sex probably occurs prior to the main event. It's the conquest. When an imagined object of desire turns into a real person, when hope is replaced with possession, the role of dopamine comes to an end. The thrill is gone, and orgasm is anticlimactic.

Finally, as would be expected when comparing liberals (with their elevated dopamine) with conservatives (with their elevated H&N neurotransmitters), conservatives are happier than liberals. A Gallup poll conducted from 2005 to 2007 found that 66 percent of Republicans were very satisfied with their lives compared to 53 percent of Democrats. Sixty-one percent of Republicans described themselves as very happy, but fewer than half of Democrats were able to say the same. In a similar vein, people who were married were happier than those who were single, and people who went to church were happier than those who did not.

The world is rarely simple, though. Despite higher rates of marital satisfaction, more reliable orgasms, and less cheating, couples in red states are more likely to get divorced than those in blue states. They also consume more pornography. Although these findings appear to be counterintuitive, one explanation is that they are the result of a greater cultural emphasis on organized religion. Red state couples are pressured to marry sooner, and they are less likely to live together or have sex prior to marriage. Consequently, the average red state couple has less opportunity to get to know each other before getting married, which may destabilize their marriage. Similarly, disapproval of premarital sex may lead to greater use of pornography to obtain sexual release.

HIPPIES AND EVANGELICALS

Adding to the complexity, political parties are heterogeneous, composed of groups who have conflicting beliefs. Among Republicans, there are the small-government advocates who believe that individuals should be left alone to make their own choices, free from controlling laws and regulations. But there are also the politically active evangelicals who want to make the country a better place by legislating morality. It's not surprising that a group that defines itself by its worship of a transcendent entity and emphasizes abstract concepts such as justice and mercy would have a more dopaminergic approach to life. Their attention to continual moral growth and the afterlife also reveals a focus on the future. They are the progressives of the right.

On the left there are the hippies who value sustainability and often frown on technology, preferring to live a life that's deeply connected to the earth. They favor the experience of the here and now over the pursuit of what they do not have. They are the conservatives of the left, rejecting the progressive arrow in favor of the conservative circle.

This complexity reminds us that when studying social trends, it's important to be careful and to maintain an open mind. The complete reversal of the results of the politics and personality traits study demonstrates that data can be mistakenly interpreted and still be accepted as correct. Even worse, the quality of data is always imperfect, and the information gathered from surveys given to thousands of people will have more errors in it than data from closely supervised clinical trials. Surveys also depend on the truthfulness of the respondents. It's possible that conservatives were less willing than liberals to admit to marital infidelity or unhappiness with life, which would have skewed the General Social Survey results.

Another problem is that scientific research can be inconsistent. Some studies on the neuroscience of political thought have an "evil twin," so to speak, that looked at the same question and found the opposite result. Overall, though, the data support a tendency toward a progressive political ideology among people with a more dopaminergic personality and a conservative one for those people with lower levels of dopamine and higher H&Ns.

The big picture might look something like this: *On average*, liberals are more likely to be forward thinking, cerebral, inconstant, creative, intelligent, and dissatisfied. Conservatives, by contrast, are more likely to be comfortable with emotions, reliable, stable, conventional, less intellectual, and happy.

THE RELIABLY IRRATIONAL VOTER

Although very conservative and very liberal people tend to vote the party line, others have more moderate ideologies. They are the independent voters who are open to political persuasion. Influencing the opinions of this group is essential for a successful campaign, and neuroscience may shed light on the best ways to do it.

The art of persuasion intersects with neuroscience at the point where decisions are made and action is taken—that is, the intersection of desire dopamine and control dopamine circuits where we weigh options and make decisions about what we think will best serve our future. Whether it's grabbing a bottle of detergent from the grocery shelf or pulling the lever for a political candidate, it looks like this should be in the realm of control dopamine, asking the simple question, *What's best for my long-term future?* But convincing control dopamine, overcoming all the counterarguments that inevitably arise, is hard to do with a bumper sticker or a 30-second television commercial. And from a purely practical point of view, it's probably not worth doing, anyway.

Rational decisions are fragile things, always open to revision as new evidence comes along. Irrationality is more enduring, and both desire dopamine and the H&N pathways can be taken advantage of to guide people toward making irrational decisions. The most effective tools are fear, desire, and sympathy.

Fear may be the most effective of them all, which is why attack ads, commercials that portray the opposing candidate as dangerous, are so popular. Fear speaks to our most primitive concerns: *Can I stay alive? Will my children be safe? Will I be able to keep my job so I'll have money for food and rent?* Stirring up fear is an indispensable part of almost any political campaign. Encouraging Americans to hate one another is an unfortunate side effect.

WHY ARE WE AMUSING OURSELVES TO DEATH?

In the provocative 1985 book *Amusing Ourselves to Death*, media scholar Neil Postman argued that political discourse was being diminished by the rise of television. He observed that television news had by then acquired many of the characteristics of entertainment. He quoted television newscaster Robert MacNeil: "The idea, he writes, 'is to keep everything brief, not to strain the attention of anyone but instead to provide constant stimulation through variety, novelty, action, and movement. You are required . . . to pay attention to no concept, no character, and no problem for more than a few seconds at a time.'" More than three decades later, news on the internet is the same way. Even outlets considered to be serious cram their home pages with dozens of brief, provocative headlines. Most lead not too long, thoughtful material for reading but to short, slick videos for watching.

Postman asserted that this poses a profound problem, but he made no guess as to why we prefer entertainment over serious thought when we debate the important questions the nation must address. Thirty years on, the question remains. Of the infinite forms communication technology might have taken, why, like TV news, has internet news elevated brevity and novelty over in-depth analysis? Aren't the events of the world worth more attention?

The answer is desire dopamine. A short, slick story stands out from the landscape—it is *salient*. It delivers a quick hit of dopamine and grabs our attention. Thus we click through a dozen provocative headlines that lead to kitten videos and skip the long essay about healthcare. The healthcare story is more pertinent to our lives, but the work of processing that story is no match for the easy pleasure of those dopamine hits. Control dopamine could push back, but it is invariably overpowered by the flood of whatever is new and shiny, and such things are the currency of the Internet.

Where will this lead? Probably not to a renaissance of long-form journalism. As quick-hit stories grow more prevalent in the news environment, they must get shorter and shallower to compete. Where does such a cycle end? Even words may not be bedrock. Most cellphones now offer to replace the text of typed phrases with something faster and simpler (and cruder) to catch the eye: an emoji.

Postman may not have known the neuroscientific cause of all this, but he understood its effect: "And so, we move rapidly into an information environment which may rightly be called trivial pursuit. As the game of that name uses facts as a source of amusement, so do our sources of news. It has been demonstrated many times that a culture can survive misinformation and false opinion. It has not yet been demonstrated whether a culture can survive if it takes the measure of the world in twenty-two minutes. Or if the value of its news is determined by the number of laughs it provides."

TO HAVE LOVED AND LOST HURTS MORE

In addition to tapping into primitive needs, another reason fear works so well is *loss aversion*, meaning that the pain of loss is stronger than the pleasure of gain. For example, the pain of losing $20 is greater than the pleasure of winning $20. That's why most people reject a 50/50 coin toss wager when the amount of money is significant. In fact, most people reject a $30 payoff for a $20 bet. The payoff must be double the wager, $40 in this case, before most people will agree to the bet.

A mathematician would say that when there is a 50/50 chance of winning, and the payoff is bigger than the bet, the gamble has a *net positive value*—you should go for it. (It's important to note that this works only if the bet is affordable. It would be rational to bet $20 you'd spend going to a movie, but not $200 you need to pay the rent.) Yet most people reject the opportunity to win $30 on a $20 coin toss. Why would they do that?

When scientists performed brain scans during wagering experiments, they naturally looked at dopamine first. They found that neural activity in the desire circuit increased after wins and decreased after losses—as would be expected. But the changes weren't symmetrical. The magnitude of the decrease after losses was larger than the increase after gains. The dopamine circuit was mirroring the subjective experience. The effect of loss was greater than the effect of gain.

What neural pathways were behind this imbalance? What was amplifying the loss reaction? The researchers turned their attention to the amygdala—an H&N structure that processes fear and other negative emotions. Every time a participant lost a bet, their amygdala fired up, intensifying feelings of distress. It was H&N emotion that was driving loss aversion. The H&N system doesn't care about the future. It doesn't care about things we might get. It cares about what we have right now. And when those things are threatened, out comes the experience of fear and distress.

Other studies found similar results. In one experiment, participants were randomly assigned to receive a coffee mug. Half the group got one, and half didn't. Immediately after handing out the mugs, the researchers gave the participants an opportunity to trade among themselves:

mugs for money. The mug owners were told to set a price they would accept, and the mug buyers were told to set a price they would pay. The mug owners asked for an average of $5.78, and the mug buyers offered an average of $2.21. The sellers were reluctant to part with their mugs. The buyers were reluctant to part with their money. Both buyers and sellers were reluctant to give up what they had.

The essential role of the amygdala in loss aversion was confirmed by something called an experiment of nature. Experiments of nature are illnesses and injuries that reveal important pieces of scientific knowledge. They are fascinating because they usually represent "experiments" that would be grossly unethical for a scientist to carry out. No one's going to ask a surgeon to open up a person's head and cut out their amygdala. But once in a while it happens on its own. In this case, scientists studied two patients who had Urbach–Wiethe disease, a rare condition that destroys the amygdala on both sides of the brain. When these individuals were presented with wagers, they attached equal weight to gain and loss. Without the amygdala, loss aversion vanished.

In a way, loss aversion is simple arithmetic. Gain is about a better future, so only dopamine is involved. The possibility of gain gets a +1 from dopamine. It gets zero from H&N, because H&N is only concerned with the present. Loss is also about the future, so it concerns dopamine, and gets a −1. Loss concerns H&N, too, because it affects things in our possession right now. So H&N gives it a −1. Put them together, and gain = +1, loss = −2, exactly what we see with the brain scans and the wagering experiments.

Fear, like desire, is primarily a future concept—dopamine's realm. But the H&N system gives a boost to the pain of loss in the form of amygdala activation, tipping our judgment when we have to make decisions about the best way to manage risk.

TO PROVIDE OR PROTECT?

Although loss aversion is a universal phenomenon, there are differences among groups. Overall, dopaminergic liberals are more likely to respond to messages that offer benefits, like opportunities for more resources, whereas H&N conservatives are more likely to respond to messages that offer security, like the ability to keep the things they currently have. Liberals support programs they believe will lead to a better future, such as subsidized education, urban planning, and government-funded technology initiatives. Conservatives prefer programs that protect their current way of life, such as defense spending, law-and-order initiatives, and limits on immigration.

Liberals and conservatives both have their reasons for focusing on threats versus benefits, reasons they believe are rational conclusions resulting from thoughtful weighing of evidence. That's probably not true. It's more likely that there is a fundamental difference in the way their brains are wired.

Researchers at the University of Nebraska selected a group of volunteers based on their political beliefs and measured their level of arousal as they were shown pictures that evoked desire or distress. Arousal is sometimes used to describe sexual excitement, but more broadly it's a measure of how engaged a person is with what's going on around him. When a person is interested and engaged, his heart beats a little faster, his blood pressure goes up a bit, and small amounts of perspiration are released from his sweat glands. Doctors call this a sympathetic response. The most common way to measure the sympathetic response is to attach electrodes to a person's body, and measure how easily electricity flows. Sweat is salt water, which conducts electricity better than dry skin. The more aroused a person is, the easier the electricity flows.

After the electrodes were attached to the research participants, they were shown three distressing photos (a spider on a man's face, an open wound with maggots, and a crowd fighting with a man) and three positive photos (a happy child, a bowl of fruit, and a cute rabbit). Liberals had a stronger response to the positive photos, conservatives to the negative ones. Because the researchers were measuring a biological reaction—perspiration—the response couldn't have been intentionally controlled by the participants. Something more fundamental than rational choice was being measured.

Next they used an eye-tracking device to determine how much time volunteers spent looking at a collage of pictures—positive and negative ones displayed at the same time. Both groups, liberal and conservative, spent more time looking at the negative pictures. This result is consistent with the universal phenomenon of loss aversion. However, the conservatives spent much more time gazing at the fear-provoking images, while the liberals divided their attention more evenly. Evidence of loss aversion was present in both groups, but it was more pronounced among conservatives.

WE HAVE WAYS OF MAKING YOU CONSERVATIVE

The relationship between conservatism and threat goes in both directions. Conservatives are more likely than liberals to focus on threat. At the same time, when people of either inclination feel threatened, they become more conservative. It's well known that terrorist attacks boost the popularity of conservative candidates. But even small threats—threats so small we're not even consciously aware of them—nudge people to the right.

To test the relationship between subtle threat and conservative ideology, researchers approached students on a college campus and asked them to fill out a survey regarding their political beliefs. Half the participants were seated in an area next to a hand sanitizer, a reminder of the risk of infection; the other half were taken to a different area. Those who sat near the hand sanitizer reported higher levels of moral, social, and fiscal conservatism. The same result occurred when a separate group of students was asked to use a germ-killing hand wipe before sitting down at a computer to answer the survey questions. It's worth noting that elections are held during flu season, and touch-screen voting machines spread germs. As a result, it's not uncommon to see hand sanitizer dispensers available for voters' use at polling places.

Professor Glenn D. Wilson, a psychologist who studies the influence of evolution on human behavior, joked that during election season, bathroom signs that say "Employees must wash hands before returning to work" are billboards for the Republican Party.

NEUROCHEMICAL MODULATION
OF MORAL JUDGMENT

Drugs work, too. Scientists can make people behave more like conservatives by giving them medication that boosts the H&N neurotransmitter serotonin. In one experiment, participants were given a single dose of the serotonergic drug citalopram, commonly used to treat depression.[3] After taking the medication, they became less focused on the abstract concept of justice and more focused on protecting individuals from harm. This was demonstrated by their performance in something called the "ultimatum game." Here's how it works.

There are two players in the ultimatum game. One player, called the proposer, is given a sum of money (e.g., $100) and told to share it with the other player, who is the responder. The proposer can offer the responder as much or as little as she likes. If the responder accepts the proposer's offer, they both keep the money. On the other hand, if the responder rejects the offer, neither player gets anything. It's a one-shot game. Each player has only one chance.

A perfectly rational responder would accept any offer, even $1. If she accepts the offer, she's financially better off than before. But if she rejects the offer, she gains nothing. Therefore, rejecting any offer, no matter how small, is contrary to her financial self-interest. Yet in reality, low offers are rejected because they offend our sense of fair play. A low offer makes us want to punish the proposer—teach her a lesson by inflicting financial

---

3   Just one dose of a serotonergic antidepressant isn't enough to influence mood. It usually takes a few weeks of daily dosing to see an effect. The first dose makes the level of serotonin in the brain go up, but after a few weeks of treatment things become more complicated. By the time the depression starts to lift, the brain has adapted to the medication in such a way that the serotonin system is more active in some places and less in others. No one really knows how antidepressants improve mood.

harm on her. On average, responders tend to punish proposers who offer 30 percent or less of the money they were told to share.

That number, 30 percent, is not fixed in stone. Different people, under different conditions, will make different decisions. Researchers at Cambridge and Harvard Universities found that participants who were given citalopram were twice as likely to accept low offers. Combining those results with the results of additional tests of moral judgment and behavior, the researchers concluded that the citalopram recipients were reluctant to harm the proposer by rejecting her offer. They found the opposite effect when they gave participants a drug that lowered serotonin levels: they were more willing to inflict harm to serve the greater good of enforcing standards of fairness.

The researchers concluded that the serotonin-boosting drug increased what they called *harm aversion.* Increasing serotonin shifts moral judgment away from an abstract goal (enforcing fairness) toward an avoidance of carrying out actions that might harm someone (depriving the proposer of her share of the money). Thinking back to the trolley problem, the logical approach is to kill one person to save five, whereas the harm-aversion approach is to refuse to take someone's life for the benefit of other people. Using drugs to influence these decisions has the unsettling name of *neurochemical modulation of moral judgment.*

The single dose of citalopram made people more willing to forgive unfair behavior and less willing to view harming another person as permissible, an attitude consistent with an H&N predominance. The researchers described this behavior as *prosocial at the individual level. Prosocial* is a term that means willingness to help other people. Rejecting unfair offers is called *prosocial at the group level.* Punishing people who make unfair offers promotes fairness that benefits the larger community, which is more consistent with a dopaminergic approach.

SHOULD THEY STAY OR SHOULD THEY GO?

We see this individual/group contrast play out in the debate over immigration. Conservatives tend to focus on smaller groups, such as individuals, family, and country, whereas liberals are more likely to focus on the largest group of all: the global community of all men and women. Conservatives are interested in individual rights, and some support the idea of building walls to keep illegal immigrants out of their country. Liberals see individuals as intertwined, and some talk about abolishing immigration laws completely. But what happens when the immigrants actually show up—when they change from an idea to a reality, from distant and abstract to right next door? There are no large-scale studies that would provide an answer to this question, but there is anecdotal evidence that the H&N experience of direct contact produces different results compared to the dopaminergic experience of setting policy.

In 2012 the *New York Times* reported on a group called Unoccupy Springs, which had arisen in the heart of the very liberal, very wealthy Hamptons. The group advocated for a crackdown on immigrants who were packing single-family homes with unrelated people in violation of the local housing code. The Unoccupy group argued that their new neighbors were overburdening the schools and depressing property values. Similarly, a study from Dartmouth College found that compared to Republican states, Democratic states have more housing restrictions that deter in-migration of lower-income people. These restrictions include limiting the number of families allowed to live in a single home and zoning restrictions that reduce the supply of affordable housing.

Harvard economist Edward Glaeser and Joseph Gyourko of the University of Pennsylvania evaluated the impact of zoning on housing affordability. They found that in most of the country, housing costs are very close to the cost of construction, but they're significantly higher in California and some East Coast cities. They note that in these areas, zoning authorities make new construction extremely costly, as much as 50 percent higher in urban areas, which are otherwise favored by immigrants.

Barriers that shut out impoverished immigrants are reminiscent of Einstein's statement, "My passionate sense of social justice and social responsibility has always contrasted oddly with my pronounced lack of need for direct contact with other human beings." Conservatives appear to be the opposite. They want to exclude illegal immigrants from this country in order to prevent what they fear will be a fundamental transformation of their culture. However, harm aversion motivates them to take care of the ones who are here.

William Sullivan, a writer for the conservative publication *American Thinker*, noted that in the midst of the debate on immigration, leading conservative figures were going to the Mexican border to assist church groups in delivering relief, including hot meals, fresh water, and a tractor trailer filled with teddy bears and soccer balls. Some called it a publicity stunt, but it's consistent with an overarching approach to life that emphasizes harm aversion: protect the status quo while protecting individuals in danger.

In opposite and complementary ways, liberals and conservatives want to help impoverished immigrants. At the same time, they both want to keep them away.

WE HAVE WAYS OF MAKING YOU LIBERAL

If introducing threats into the environment makes people more conservative, is it possible to make people more liberal by doing the opposite? Dr. Jaime Napier, an expert on political and religious ideologies, found that the answer is yes, and it doesn't take very much prodding. Just as researchers were able to increase conservatism with the tiny nudge of putting a hand sanitizer nearby, Dr. Napier was able to make people more liberal with a simple imagination exercise. She told conservatives to imagine they had superpowers that made it impossible for them to be injured. Subsequent testing of political ideology found that they became more liberal. Reducing feelings of vulnerability, which subsequently suppressed H&N fear of loss, allowed dopamine, the agent of change, to switch on and play a larger role in determining ideology.

What about the act of imagining all by itself? Imagining is a dopaminergic activity because it involves things that have no physical existence. Did simple activation of the dopamine system through the exercise of imagination contribute to the leftward shift in political beliefs? A separate study suggests that it did.

Abstract thinking is one of the primary functions of the dopamine system. Abstract thinking allows us to go beyond sensory observation of events to construct a model that explains why the events are occurring. A description that relies on the senses focuses on the physical world: things that actually exist. The technical term for this type of thinking is *concrete*. That's an H&N function, and scientists call it *low-level* thinking. Abstract thinking is called *high-level*. A group of scientists wondered if people who tended to think concretely would be more hostile to groups who were different from themselves—people they perceived as threatening the stability of their way of life—such as gays, lesbians, Muslims, and atheists.

Research volunteers were given two descriptions of events, such as ringing a doorbell. They had to choose which description was best. One was concrete (ringing a doorbell is moving a finger), and the other was abstract (ringing a doorbell is seeing if anyone is home). Next they asked them to rate their feelings of liking and warmth for gays, lesbians, Muslims, and atheists. They found a direct relationship between choosing concrete descriptions and reporting lower levels of liking and warmth.

The next step was to see if these feelings could be manipulated by stimulating the participants to think abstractly. They chose the subject of exercise, a topic completely unrelated to the acceptance of potentially threatening groups. The researchers began by asking the participants to think about maintaining good physical health. Then half were asked to describe *how* they would do it (concrete) and the other half were asked to describe *why* it's important (abstract). Describing *how* had no effect on attitudes, but describing *why* raised the conservative participants' feelings of liking and warmth for the unfamiliar groups to the point where there was no significant difference between their attitudes and the attitudes of liberals.

Activating dopamine pathways is one way to make conservatives think more like liberals. But we can do something similar by exploiting the very circuits that make conservatives act conservatively: the H&N

circuits, specifically those that allow us to experience empathy. This approach uses strengths that are quintessentially conservative to generate greater acceptance of people who threaten change.

Let's go back to the apparent contradiction of conservatives advocating for deportation of illegal immigrants as a group while providing individuals with food, water, and toys. H&N conservatives may be hostile to the idea of immigration, but they have an innate ability to connect on an empathic basis to actual immigrants. This ability—one might even call it an unconscious impulse—has been used by Hollywood writers to increase acceptance of lesbian, gay, bisexual, and transgender (LGBT) people. They do it through the power of story.

We develop emotional relationships with characters in stories. If it's a well-written story, the feelings we have for the characters may be very similar to the feelings we have for real people. The Gay & Lesbian Alliance Against Defamation (GLAAD) notes, "TV hasn't merely reflected the changes in social attitudes; it has also had an important role in bringing them about. Time and again, it's been shown that personally knowing an LGBT person is one of the most influential factors in shifting one's views on LGBT issues, but in the absence of that, many viewers have first gotten to know us as television characters."

According to GLAAD's annual report on the diversity of primetime TV, the number of regular characters identified as gay, lesbian, or bisexual has been steadily growing. In the most recent poll, conducted in 2015, it was 4 percent. That's about the same as the 3.8 percent of Americans who identified as LGBT in a recent Gallup poll. The network with the highest percentage was Fox Network, where 6.5 percent of regular primetime characters were LGBT.

These fictional characters have a real influence on viewer attitudes. A poll conducted by *The Hollywood Reporter* found that 27 percent of respondents said LGBT-inclusive TV made them stronger advocates for same-sex marriage. When the results were analyzed based on how viewers voted in the 2012 presidential election, 13 percent of Romney voters said that watching the shows made them more supportive of same-sex marriage. Transforming abstract groups into concrete individuals is a good way to activate H&N empathy circuits.

A NATION GOVERNED BY IDEAS (VIA BIOLOGY)

[According to] Ashleymadison.com, a dating Web site for married people looking for extramarital affairs, . . . [Washington, DC] topped a list ranking the country's most adulterous cities for the third year in a row . . . And the neighborhood with the most cheaters? Capitol Hill, the land of politicians, staffers and lobbyists.
—Washington Post, May 20, 2015

The essence of government is control. People may submit to being controlled as a result of conquest, or they may voluntarily give up some of their freedom in exchange for protection. Either way, a small number of people are given power to exert authority over the rest of the population. It's a dopaminergic activity because the populace is governed from a distance through abstract laws. Although the threat of H&N violence is used to enforce the law, most people never experience it. They submit to ideas, not physical force.

Since government is inherently dopaminergic, liberals tend to be more enthusiastic about it than H&N conservatives. Five hundred liberals marching down the street are probably staging a protest. With conservatives, it's more likely a parade. In addition to their enthusiasm for engaging in the political process, liberals are also more likely to pursue advanced degrees in public policy, and they're often attracted to fields such as journalism in which they are involved in the political process on a daily basis. Conservatives, by contrast, are more often distrustful of government, especially government that acts at a distance. Conservatives tend to prefer local governance, with power exerted at the state or local level rather than federally.

Distance matters. Thinking back to the trolley problem, it's easier to maximize resources when emotions are taken out of the picture. Pushing a person onto the tracks to stop a train is nearly impossible. Flipping a switch from far away is easier. Similarly, many laws benefit some people but harm others. The farther away you get, the easier it is to tolerate some degree of harm in the service of the greater good. Distance insulates politicians from the immediate consequences of their decisions. Raise a tax, cut funding, send someone to war; the person taking home less pay, receiving less help, or hunkered down in a foxhole will rarely be in the company of the person who put him in that position, as long as that person is in Washington, DC. There's no opportunity for H&N circuits to trigger distressing emotions that would make these decisions more difficult.

WHY WASHINGTON MUST ALWAYS "DO SOMETHING!"

Apart from distance, another way in which government is fundamentally dopaminergic is that it is about *doing something*. It's almost unheard of for a politician to campaign on a promise that he will go to Washington and do nothing. Politics is about change and change is driven by dopamine. Whenever tragedy strikes, the cry goes up, *Do something!* So airport security is beefed up after a terrorist attack despite evidence that the long, humiliating rituals travelers must endure don't really increase safety. Undercover TSA agents who test the system can almost always get weapons through. Nevertheless, the mandate to *do something* gets fulfilled.

According to GovTrack.us, the federal government enacted between 200 and 800 laws during each two-year congressional session since 1973. That's a lot of laws, but it's nothing compared to what politicians tried to do. During these sessions, Congress made attempts to pass between 8,000 and 26,000 laws. When the people think something ought to be done, politicians are happy to oblige.

This desire for control is unavoidable. Some people in Washington call themselves liberal and others call themselves conservative, but pretty much everyone involved in politics is dopaminergic. Otherwise they couldn't get elected. Political campaigns require intense motivation. They require a willingness to sacrifice everything to achieve success. Long hours take a toll on family life in particular. H&N people, who make relationships with loved ones a priority, can't succeed in politics. In the United Kingdom, the divorce rate among members of Parliament is double that of the general population. In the United States it's common for members of Congress to live in Washington, while their families live back in their home states. They rarely see their spouses, and there are plenty of young staffers enamored with power that are available to satisfy dopaminergic desires. To a politician, relationships aren't for enjoyment; they're for a purpose, whether it's to get elected, pass a bill, or satisfy a biological urge. As President Harry Truman is credited with saying, "If you want a friend in Washington, buy a dog."

CONSERVATIVE CANDIDATE, LIBERAL LAWMAKER

The need to be dopaminergic in order to get elected is a problem for conservatives because having dopaminergic politicians represent H&N constituents doesn't always work well. In recent years conservatives have experienced growing frustration with so-called establishment Republicans who campaign on promises to scale back the government but end up growing it instead. The Tea Party is the most visible manifestation of this frustration. This conservative group generated unusual enthusiasm, yet so far it has been unable to achieve its goal to slow the growth of government.

Growth may never stop. Dopamine's mandate is *more*. Change—representing either progress or loss of tradition, depending on one's point of view—is inevitable. Only H&N circuits can bring about feelings of satisfaction, feelings that the end has been reached and it's time to stop. Endorphins, endocannabinoids, and other H&N neurotransmitters tell us that our work is done, and now it's time to enjoy the fruits of our labor. But dopamine suppresses these chemicals. Dopamine never rests. The game of politics is played 24 hours a day, seven days a week, and stopping to take a breath or saying the word "enough" leads to failure.

That's not to say that more government is necessarily bad. Growth of power wielded for the public good can have a positive influence on the lives of millions of people. If the government is benevolent and effective, growing centralized power can help safeguard the rights of the weak and lift the destitute out of poverty. It can protect workers and consumers from exploitation by powerful corporations. But if politicians pass laws that benefit themselves instead of their constituents, if corruption is widespread, or if lawmakers simply don't know what they are doing, liberty and prosperity will suffer.

Historically, the only way to reverse the expansion of power is to replace incremental change with cataclysmic change in the form of revolution. John Calhoun, the nineteenth-century South Carolina senator and vice president, showed an understanding of the type of person who plays the game of power—whether the player is a rebel or a tyrant—when he said that it is easier to obtain liberty than it is to preserve it. Rebels are dopaminergic and politicians are dopaminergic. The goal of both is change.

DON'T GET FOOLED AGAIN

In the end, the fundamental obstacle to achieving harmony is that the liberal brain is different from the conservative brain, and that makes it difficult for them to understand each other. Because politics is an adversarial game, this lack of understanding leads to demonization of the other side. Liberals believe conservatives want to take the country back to a time when minorities were treated with gross injustice. Conservatives believe liberals want to pass repressive laws that control every aspect of their lives.

In reality, the vast majority of people on both sides of the political divide want what's best for all Americans. There are exceptions; there are bad people everywhere, and it's the bad people who get all the press. They're more interesting than good people, and they're useful as political weapons. But they don't represent the typical Democrat or Republican.

Most conservatives just want to be left alone. They want the freedom to make their own decisions based on their own values. Most liberals want to help people live better lives. Their goal is for everyone to be healthier, safer, and free from discrimination. But political leaders bene-fit from stirring up hostility between the two groups because it strength-ens the allegiance of their followers. The important thing to remember is that liberals want to help people become better, conservatives want to let people be happy, and politicians want power.

FURTHER READING

Verhulst, B., Eaves, L. J., & Hatemi, P. K. (2012). Correlation not causation: The relationship between personality traits and political ideologies. *American Journal of Political Science, 56*(1), 34–51.

Bai, M. (2017, June 29). Why Pelosi should go—and take the '60s generation with her. *Matt Bai's Political World.* Retrieved from www.yahoo.com/news/pelosi -go-take-60s-generation-090032524.html

Gray, N. S., Pickering, A. D., & Gray, J. A. (1994). Psychoticism and dopamine D2 binding in the basal ganglia using single photon emission tomography. *Personality and Individual Differences, 17*(3), 431–434.

Eysenck, H. J. (1993). Creativity and personality: Suggestions for a theory. *Psychological Inquiry, 4*(3), 147–178.

Ferenstein, G. (2015, November 8). Silicon Valley represents an entirely new political category. TechCrunch. Retrieved from https://techcrunch.com/2015/11 /08/silicon-valley-represents-an-entirely-new-political-category/

Moody, C. (2017, February 20). Political views behind the 2015 Oscar nominees. CNN. Retrieved from http://www.cnn.com/2015/02/20/politics/ oscars-political-donations-crowdpac/

Robb, A. E., Due, C., & Venning, A. (2016, June 16). Exploring psychological wellbeing in a sample of Australian actors. *Australian Psychologist.*

Wilson, M. R. (2010, August 23). Not just News Corp.: Media companies have long made political donations. *OpenSecrets Blog.* Retrieved from https://www. opensecrets.org/news/2010/08/news-corps-million-dollar-donation/

Kristof, N. (2016, May 7). A confession of liberal intolerance. *The New York Times.* Retrieved from http://www.nytimes.com/2016/05/08/opinion/sunday /a-confession-of-liberal-intolerance.html

Flanagan, C. (2015, September). That's not funny! Today's college students can't seem to take a joke. *The Atlantic.*

Kanazawa, S. (2010). Why liberals and atheists are more intelligent. *Social Psychology Quarterly, 73*(1), 33–57.

Amodio, D. M., Jost, J. T., Master, S. L., & Yee, C. M. (2007). Neurocognitive correlates of liberalism and conservatism. *Nature Neuroscience, 10*(10), 1246–1247.

Settle, J. E., Dawes, C. T., Christakis, N. A., & Fowler, J. H. (2010). Friendships moderate an association between a dopamine gene variant and political ideology. *The Journal of Politics*, *72*(4), 1189–1198.

Ebstein, R. P., Monakhov, M. V., Lu, Y., Jiang, Y., San Lai, P., & Chew, S. H. (2015, August). Association between the dopamine D4 receptor gene exon III variable number of tandem repeats and political attitudes in female Han Chinese. *Proceedings of the Royal Society B*, *282*(1813), 20151360.

How states compare and how they voted in the 2012 election. (2014, October 5). *The Chronicle of Philanthropy*. Retrieved from https://www.philanthropy.com/article/How-States-CompareHow/152501

Giving USA. (2012). *The annual report on philanthropy for the year 2011*. Chicago: Author.

Kertscher, T. (2017, December 30). Anti-poverty spending could give poor $22,000 checks, Rep. Paul Ryan says. Politifact. Retrieved from http://www.politifact.com/wisconsin/statements/2012/dec/30/paul-ryan/anti-poverty-spending-could-give-poor-22000-checks/

Giving USA. (2017, June 29). Giving USA: Americans donated an estimated $358.38 billion to charity in 2014; highest total in report's 60-year history [Press release]. Retrieved from https://givingusa.org/giving-usa-2015-press-release-giving-usa-americans-donated-an-estimated-358-38-billion-to-charity-in-2014-highest-total-in-reports-60-year-history/

Konow, J., & Earley, J. (2008). The hedonistic paradox: Is homo economicus happier? *Journal of Public Economics*, *92*(1), 1–33.

Post, S. G. (2005). Altruism, happiness, and health: It's good to be good. *International Journal of Behavioral Medicine*, *12*(2), 66–77.

Brooks, A. (2006). *Who really cares?: The surprising truth about compassionate conservatism*. Basic Books.

Leonhardt, D., & Quealy, K. (2015, May 15). How your hometown affects your chances of marriage. *The Upshot* [Blog post]. Retrieved from https://www.nytimes.com/interactive/2015/05/15/upshot/the-places-that-discourage-marriage-most.html

Kanazawa, S. (2017). Why are liberals twice as likely to cheat as conservatives? *Big Think*. Retrieved from http://hardwick.fi/E%20pur%20si%20muove/why-are-liberals-twice-as-likely-to-cheat-as-conservatives.html

Match.com. (2012). Match.com presents Singles in America 2012. *Up to Date* [blog]. Retrieved from http://blog.match.com/sia/

Dunne, C. (2016, July 14). Liberal artists don't need orgasms, and other findings from OkCupid. Hyperallergic. Retrieved from http://hyperallergic.com /311029/liberal-artists-dont-need-orgasms-and-other-findings-from-okcupid/

Carroll, J. (2007, December 31). Most Americans "very satisfied" with their personal lives. Gallup.com. Retrieved from http://www.gallup.com/ poll/103483/most-americans-very-satisfied-their-personal-lives.aspx

Cahn, N., & Carbone, J. (2010). *Red families v. blue families: Legal polarization and the creation of culture.* Oxford: Oxford University Press.

Edelman, B. (2009). Red light states: Who buys online adult entertainment? *Journal of Economic Perspectives, 23*(1), 209–220.

Schittenhelm, C. (2016). What is loss aversion? *Scientific American Mind, 27*(4), 72–73.

Kahneman, D., Knetsch, J. L., & Thaler, R. H. (1991). Anomalies: The endowment effect, loss aversion, and status quo bias. *Journal of Economic Perspectives, 5*(1), 193–206.

De Martino, B., Camerer, C. F., & Adolphs, R. (2010). Amygdala damage eliminates monetary loss aversion. *Proceedings of the National Academy of Sciences, 107*(8), 3788–3792.

Dodd, M. D., Balzer, A., Jacobs, C. M., Gruszczynski, M. W., Smith, K. B., & Hibbing, J. R. (2012). The political left rolls with the good and the political right confronts the bad: Connecting physiology and cognition to preferences. *Philosophical Transactions of the Royal Society B: Biological Sciences, 367*(1589), 640–649.

Helzer, E. G., & Pizarro, D. A. (2011). Dirty liberals! Reminders of physical cleanliness influence moral and political attitudes. *Psychological Science, 22*(4), 517–522.

Crockett, M. J., Clark, L., Hauser, M. D., & Robbins, T. W. (2010). Serotonin selectively influences moral judgment and behavior through effects on harm aversion. *Proceedings of the National Academy of Sciences, 107*(40), 17433–17438.

Harris, E. (2012, July 2). Tension for East Hampton as immigrants stream in. *The New York Times.* Retrieved from http://www.nytimes.com/2012/07/03 /nyregion/east-hampton-chafes-under-influx-of-immigrants.html

Glaeser, E. L., & Gyourko, J. (2002). *The impact of zoning on housing affordability* (Working Paper No. 8835). Cambridge, MA: National Bureau of Economic Research.

Real Clear Politics. (2014, July 9). Glenn Beck: I'm bringing soccer balls, teddy bears to illegals at the border. Retrieved from http://www.realclearpolitics.com/video/2014/07/09/glenn_beck_im_bringing_soccer_balls_teddy_bears_to_illegals_at_the_border.html

Laber-Warren, E. (2012, August 2). Unconscious reactions separate liberals and conservatives. *Scientific American*. Retrieved from http://www.scientificamerican.com/article/calling-truce-political-wars/

Luguri, J. B., Napier, J. L., & Dovidio, J. F. (2012). Reconstruing intolerance: Abstract thinking reduces conservatives' prejudice against nonnormative groups. *Psychological Science, 23*(7), 756–763.

GLAAD. (2013). *2013 Network Responsibility Index*. Retrieved from http://glaad.org/nri2013

GovTrack. (n.d.). Statistics and historical comparison. Retrieved from https://www.govtrack.us/congress/bills/statistics

. . . the beginning is where the end gets born.
—Catherynne M. Valente, writer

Chapter 6
PROGRESS

What happens when the servant becomes the master?

*In which dopamine ensures the survival of early humans
and the extinction of the human race.*

OUT OF AFRICA

Modern humans evolved in Africa about 200,000 years ago and began spreading to other parts of the world approximately 100,000 years later. This migration was essential for the survival of the human race, and there's genetic evidence that we almost didn't make it. One of the unusual characteristics of the human genome is that there is far less variation from person to person compared to other primate species such as chimpanzees or gorillas. This high level of genetic similarity suggests that we are all descendants of a relatively small number of ancestors. In fact, early in our evolutionary history, unknown events killed off a large portion of humans, and the population dwindled to less than 20,000, representing a serious risk of extinction.

That near-extinction event illustrates why migration is so important. When a species is concentrated in a small area, there are many ways in which the entire population can be wiped out. Drought, disease, and other disasters can easily cause extinction. Dispersing throughout many regions, on the other hand, is like an insurance policy. The destruction of one population doesn't result in total extinction.

Based on the appearance and frequency of genetic markers in modern peoples, scientists estimate that early humans spread out across Asia about 75,000 years ago. They reached Australia 46,000 years ago and made it to Europe 43,000 years ago. Migration to North America occurred later, sometime between 30,000 and 14,000 years ago. Today, humans occupy nearly every corner of the globe—but not because humans recognized the threat and dispersed.

THE ADVENTURE GENE

Research on mice has shown that drugs that boost dopamine also increase exploratory behavior. Mice given these drugs move around their cages more and are less timid about entering unfamiliar environments. So could dopamine have helped propel early humans out of Africa and across the globe? To answer this question, scientists from the University of California compiled data from twelve studies that measured the frequency of dopaminergic genes in different parts of the world.

They focused on the gene that tells the body how to make the D4 dopamine receptor (*DRD4*). You may recall that dopamine receptors are proteins that are attached to the outside of brain cells. A dopamine receptor's job is to wait for a dopamine molecule to come along and bind to it. Binding sets off a cascade of chemical reactions inside the cell that changes the way the cell behaves.

We encountered this gene before when we described the connection between novelty-seeking and political ideology. Recall that genes come in different varieties called alleles. Alleles represent slight variations in the coding of genes that give people different characteristics. People who have a long form of the *DRD4* gene, such as the 7R allele, are more likely to take risks. They pursue new experiences because they have a low tolerance for boredom. They like to explore new places, ideas, foods, drugs, and sexual opportunities. They are adventurers. Worldwide about one in five people have the 7R allele, but there's substantial variation from place to place.

MORE DOPAMINE, MORE DISTANCE

The researchers obtained genetic data from the most well-known migration routes in North America, South America, East Asia, Southeast Asia, Africa, and Europe. When they analyzed the data, a clear pattern emerged. Among populations that remained near their origins, fewer people had a long *DRD4* allele compared to those who migrated farther away.

One of the migration routes they evaluated began in Africa, went through East Asia, across the Bering Strait to North America, then down to South America. That's a long way—and the researchers found that the group that made it all the way, indigenous South Americans, had the highest proportion of long dopamine alleles, 69 percent. Among those who migrated a shorter distance and settled in North America, only 32 percent had the long allele. Indigenous populations in Central America were right in between at 42 percent. On average, it was estimated that the proportion of long alleles increased by 4.3 percentage points for every 1,000 miles of migration.

Once it was established that the 7R allele of the *DRD4* gene was related to how far a population migrated, the next question was why? How did the 7R allele become more common in far-flung populations? The obvious answer is that dopamine makes people seek out more. It makes them restless and dissatisfied. It makes them long for something better. These are exactly the kinds of people who would leave an established community to go out and explore the unknown. But there's another explanation as well.

SURVIVAL OF THE FITTEST

Maybe the migratory tribes left for some other reason that had nothing to do with novelty-seeking. Maybe they left because of conflict, or perhaps they were hunting migratory animals. There could have been many reasons unrelated to dopamine, but the question remains: Under these circumstances, why would the migratory population end up with lots of 7R alleles among its members? The answer is that maybe the 7R allele didn't set off the migration, but once it began, the allele gave its carriers a survival advantage.

One advantage provided by the 7R allele is that it drove its carriers to explore the new environment in which they found themselves in order to seek out opportunities to maximize resources. In other words, it promoted novelty-seeking. For example, a tribe might have started out in a geographical area where there was a consistent climate, and the same type of food was available all year round. However, after moving to a new location, the members of the migratory tribe may have experienced rainy and dry seasons, and they needed to learn how to switch food sources as the seasons changed. Figuring out how to do that involved risk-taking and experimentation.

There's also evidence that people who carry the 7R allele are faster learners, especially when getting the answer right triggers a reward. In general, 7R carriers are more sensitive to rewards; they have stronger reactions to both wins and losses. Consequently, when they found themselves in an unfamiliar environment and needed to adapt to new routines to stay alive, the 7R carriers worked harder to figure things out because their experiences of success and failure were more intense.

Another advantage is that the 7R allele is associated with something called *low reactivity to novel stressors*. Change is stressful—both good change and bad change. For example, there are few things more stressful than divorce, but getting married is hard, too. Going bankrupt is stressful but so is winning the lottery. Bad changes may cause more stress than good changes, but the most important factor is the size of the change. Bigger change means more stress.

Stress isn't good for human health. In fact, stress kills. Stress increases the likelihood of developing heart disease, poor sleep, digestive problems, and immune system impairment. It can also trigger depression, which leads to low energy, poor motivation, hopelessness, thoughts of death, and simply giving up, all of which militate against survival. Among our evolutionary ancestors, people who were sensitive to stress had a harder time extracting resources from environments that represented a large change from what they were used to. They were less successful hunters and less productive gatherers. That made it hard for them to compete for reproductive mates, and sometimes they didn't even live long enough to have children who would carry their genes forward to the next generation.

Not everyone gets stressed by change, though. A new job, a new city, even a whole new career can be exciting and energizing for people with dopaminergic personalities. They thrive in unfamiliar environments. In prehistoric times, they were more likely to cope well despite radical changes in their way of life. They competed more successfully for mates, and as a result they passed along their dopaminergic genes. Over time, alleles that helped people adjust to unfamiliar environments with ease became more common in the population, while other alleles became rarer.

Of course, carriers of the 7R allele weren't well suited to every environment. People with dopaminergic personalities may do well when coping with novel situations, but they often have difficulty with relationships. That's important because skillful social functioning also provides an evolutionary advantage. No matter how big, how strong, or how smart a person is, he's not going to be able to compete with people who work together as a group. Individuals shouldn't fight gangs. In this situation, when the need for cooperation is paramount, a dopaminergic personality is a liability.

So it all depends on the environment. Under familiar conditions, in which social cooperation counts the most, highly dopaminergic genes become less common because their survival and mate-seeking advantages diminish relative to the benefits of more balanced dopamine levels. On the other hand, when a tribe picks up and heads off into the unknown, genes that give a person a more active dopamine system provide an advantage and become more common over time.

WHICH IS RIGHT?

That leaves us with two competing theories:

1. Dopaminergic genes propelled people to seek new opportunities. As a result these genes are found more frequently among populations who migrated from their evolutionary origins.
2. Something else made them seek new opportunities, and the dopaminergic genes allowed some of them to survive and reproduce more successfully than others.

How do we decide which one is correct?

This is where it gets a little complicated. If dopaminergic genes got people started (i.e., set them off to seek a better life), then we should see lots of 7R alleles in every group that left Africa. That would be the case whether they migrated for a few generations and ended up close to their origin, or migrated for many generations and ended up far away. That's because if it takes lots of dopamine to get started, where the tribe ended up shouldn't matter. Those who left would have a lot, and those who stayed would have less.

On the other hand, if people got started without the need for the 7R allele, then we'd see a more gradual change in the number of people who carry it. Here's why. If a tribe migrated only a short distance, only a few generations would experience unfamiliar environments. Once they stopped moving, the unknown became the familiar, and the 7R allele no longer conferred an advantage. Once the playing field was level, the 7R allele carriers lost the ability to have more children than their less dopaminergic neighbors. At this point, all the different alleles were passed along equally to subsequent generations of offspring.

Tribes that kept going, however, would experience unfamiliar environments generation after generation after generation. The reproductive advantages of 7R would continue, and 7R carriers would live longer and have more children. Over time the 7R allele would become more and more common among these long-distance travelers. And that's what we do see. The farther a population migrated, the greater the frequency of the 7R allele. It didn't start them moving, but it did help them survive as they went along.

IMMIGRATION

Movement across the globe today is different from what our prehistoric ancestors experienced. Emigration away from one's native country is a personal decision rather than a tribal decision. And although the reason may be similar—seeking better opportunities—the 7R allele of the D4 dopamine receptor doesn't seem to play a role. Immigrant populations have about the same percentage of the 7R allele as the people who remained in their home country. Nevertheless, dopamine seems to be involved, but in a different way.

In chapter four, when we discussed the role of dopamine in creativity, we compared creativity to schizophrenia, a mental illness characterized by excessive dopamine in the desire circuit. We discussed ways in which psychotic delusions have things in common with highly creative ideas and even ordinary dreams. But schizophrenia is not the only illness characterized by excessive dopamine activity. Bipolar disorder, sometimes called manic-depressive illness, also has a dopaminergic component, and the condition seems to be linked to immigration.

BIPOLAR MANIA: ANOTHER CONDITION
OF TOO MUCH DOPAMINE

Bipolar refers to two mood extremes (like *bicycle* refers to two wheels). People with bipolar disorder experience episodes of depression when their mood is abnormally low and episodes of mania when it's too high. The latter is associated with high levels of dopamine, which shouldn't be surprising given the symptoms of the manic state: high energy, euphoric mood, racing thoughts that quickly jump from one topic to another, an abundance of activity in pursuit of many goals at once, and excessive involvement in high-risk, pleasure-seeking activities such as unrestrained spending and promiscuous sexual behavior.

Many people with bipolar disorder are disabled by the illness. They are unable to hold down a job or maintain healthy relationships. Others, typically those receiving medical treatment, are able to live normal lives while taking mood stabilizing medication. A few live extraordinary lives. Worldwide, about 2.4 percent of the population has bipolar disorder, but it is more common among certain groups. Researchers in Iceland found that people who worked in creative fields such as dancing, acting, music, and writing were about 25 percent more likely to have bipolar disorder compared to those with noncreative jobs. In another study, scientists from the University of Glasgow followed over 1,800 individuals from the age of eight to their early twenties. They found that higher IQ scores at age eight predicted greater risk of developing bipolar disorder by the age of 23. Smarter brains had a greater risk of developing a dopaminergic mental illness compared to ordinary ones.

Many famous, creative people have revealed that they live with bipolar disorder. Among them are Francis Ford Coppola, Ray Davies, Patty Duke, Carrie Fisher, Mel Gibson, Ernest Hemingway, Abbie Hoffman, Patrick Kennedy, Ada Lovelace, Marilyn Monroe, Sinéad O'Connor, Lou Reed, Frank Sinatra, Britney Spears, Ted Turner, Jean-Claude Van Damme, Virginia Woolf, and Catherine Zeta-Jones. There are also many noteworthy people from the past who, based on historical documents, are thought to have had bipolar disorder. They include Charles Dickens, Florence Nightingale, Friedrich Nietzsche, and Edgar Allan Poe.

One might conceptualize the extraordinary brain as being similar to a high-performance sports car. It's capable of doing incredible things, but it breaks down easily. Dopamine drives intelligence, creativity, and hard work, but it can also make people behave in bizarre ways.

Excessive dopamine activity isn't the only problem in bipolar mania, but it plays an important role. As noted, it's not caused by a highly active *DRD4* receptor allele. Instead, scientists believe that it's caused by a problem with something called *the dopamine transporter* (Figure 5).

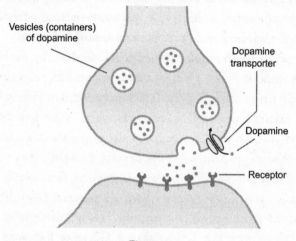

Figure 5

The dopamine transporter is like a vacuum cleaner. Its job is to limit the amount of time dopamine spends stimulating the cells around it. When a dopamine-producing cell fires, it releases its store of dopamine, which then binds to receptors on other brain cells. Then, to bring the interaction to an end, the dopamine transporter sucks the dopamine back into the cell where it came from so the process can start all over again. The transporter is sometimes called a *reuptake pump* because it pumps the dopamine back into the cell.

What happens when the transporter doesn't function normally? We can answer this question by looking at the behavior of people who abuse cocaine. Cocaine blocks the dopamine transporter like a sock shoved into a vacuum cleaner nozzle. The blockage allows the dopamine to interact with its receptor over and over again. When that happens, people experience increased energy, goal-directed activity, and sexual drive. They have elevated self-esteem, euphoria, and racing thoughts that jump from one topic to another. Cocaine intoxication is so similar to mania that doctors have trouble telling them apart.

DO BIPOLAR GENES DRIVE IMMIGRATION?

I learned very quickly that when you emigrate, you lose the crutches that have been your support; you must begin from zero, because the past is erased with a single stroke and no one cares where you're from or what you did before.
—Isabel Allende, writer

Bipolar disorder isn't all or nothing. Some people have severe forms of the illness and others have more mild forms. Some people have only a bipolar tendency. We can see things in the personalities of this latter group that suggest unusually elevated moods, but not so bad that we would diagnose them with a medical illness. It all depends on how many risk genes a person inherits from her parents, and how much vulnerability these genes confer. The genetic risk then interacts with a person's environment (a stressful childhood, for example), and the final product is some manifestation of bipolar disorder, or bipolar characteristics not severe enough to cause the actual illness.

Is it possible that minor dysfunction in the dopamine transporter—just a few risk genes or genes that have only a mild effect—could give people "itchy feet," so to speak? Might that play a role in the decision to leave one's home and seek new opportunities in a foreign country? It's not easy to pull up one's roots, to say goodbye to friends and family, and leave a community that's familiar, comfortable, and supportive. Andrew Carnegie, a nineteenth-century Scottish immigrant who started working in a factory for pennies a day and later became the richest man in the world, wrote, "[the] contented do not brave the waves of the stormy Atlantic, but sit helplessly at home."

If bipolar genes promote emigration, these ambitious people would carry their risk genes with them, and we would expect to find high concentrations of bipolar genes in countries that have lots of immigrants. The United States is populated almost entirely by immigrants and their descendants. It also has the highest rate of bipolar disorder: 4.4 percent, which is about twice the rate of the rest of the world. Are the two related?

Japan, which has almost no immigration, has a bipolar rate of 0.7 percent, one of the lowest in the world. People in the United States with bipolar disorder also start to have symptoms at a younger age, a marker of a more severe form of the illness. About two-thirds develop symptoms before the age of 20, compared to only a quarter in Europe. That supports the idea that the gene pool in the United States has a greater concentration of high-risk genes.

The gene that tells the body how to make the dopamine transporter is one of these genes, but there are many others. No one knows precisely how many, but it's clear that some form of genetic inheritance is playing a role. Children of bipolar parents are at least twice as likely to develop bipolar disorder compared to the general population. Some studies have found the risk to be ten times as high. But sometimes the children get lucky. They get the advantages bipolar people enjoy without getting the illness itself.

As noted, bipolar disorder isn't all or nothing. Mood-disorder specialists talk about a *bipolar spectrum*. At one end of the spectrum is bipolar I. People with this form of the illness experience severe mania and severe depression. Next comes bipolar II. People with bipolar II experience severe depression, but more mild episodes of elevated mood called hypomania (*hypo* means below, like a hypodermic injection that delivers a drug underneath the skin). Farther down the spectrum is cyclothymia, which is characterized by cycles of hypomania and mild depressive episodes. Then there is something called hyperthymic temperament, derived from the Greek word *thymia*, which means state of mind.

Hyperthymic temperament is not considered an illness. It doesn't occur in episodes like bipolar disorder. People with hyperthymic temperament just have a "hyper" personality, and they have it all the time. According to Hagop Akiskal, who did much of the pioneering work in this area, people with hyperthymic temperaments are upbeat, exuberant, jocular, overoptimistic, overconfident, boastful, and full of energy and plans. They are versatile with broad interests, overinvolved and meddlesome, uninhibited and risk-taking, and they generally don't sleep very much. They become overly enthusiastic about new directions in their lives, such as diets, romantic partners, business opportunities, even religions, and then quickly lose interest. They often accomplish a great deal, but they can be difficult to live with.

The last stage of the bipolar spectrum belongs to people who inherit a very limited amount of genetic risk. These people don't experience any abnormal symptoms, but they do enjoy such things as enhanced motivation, creativity, a tendency toward bold action and risk-taking, and other characteristics that reflect higher than average levels of dopamine activity.

DOPAMINE NATION

We find bipolar genes and bipolar disorder in a relatively high concentration in the United States. What about those non-pathological expressions of the condition? Is there any evidence that these conditions are also widespread? In fact, the evidence is abundant, going all the way back to the early years of the republic.

One of the earliest observers of American culture was Alexis de Tocqueville, a French diplomat, political scientist, and historian. Tocqueville described his observations of the character of Americans during the nineteenth century in his book *Democracy in America*. He studied the new country because he believed that democracy was likely to supplant aristocracy back in Europe. He thought that a study of the effects of democracy in the United States would be useful to Europeans as they navigated new forms of government.

Much of what Tocqueville observed could be attributed to the democratic principle of egalitarianism. But he also described characteristics of Americans that did not seem to be related to political philosophy. Some of these characteristics are strikingly similar to the symptoms of bipolar disorder, or at least a dopaminergic personality. For example, he devotes a chapter to "Fanatical Enthusiasm in Some Americans." He wrote,

> *Although the desire of acquiring the good things of this world is the prevailing passion of the American people, certain momentary outbreaks occur, when their souls seem suddenly to burst the bonds of matter by which they are restrained, and to soar impetuously towards heaven.*

In this single sentence we see the passionate pursuit of *more* as well as an attraction to things beyond the realm of the physical senses—even a reference to the extrapersonal space of up, the realm of heaven. Tocqueville found that behaviors of this nature were particularly common "in the half-peopled country of the Far West," a notion consistent with the likelihood that the adventurous pioneers who settled the western states were more likely to have risk-taking, sensation-seeking personalities, and possibly genetic loading for hyperdopaminergic states.

A subsequent chapter titled "Causes of the Restless Spirit of Americans in the Midst of Their Prosperity" expanded on the dopaminergic theme of never enough. Tocqueville noted that despite living in "the happiest circumstances which the world affords," Americans pursued a better life with "feverish ardor." He wrote:

> *In the United States a man builds a house to spend his later years in, and he sells it before the roof is on: he plants a garden, and rents it just as the trees are coming into bearing: he brings a field into tillage, and leaves other men to gather the crops: he embraces a profession, and gives it up: he settles in a place, which he soon afterwards leaves, to carry his changeable belongings elsewhere. If his private affairs leave him any leisure, he instantly plunges into the vortex of politics; and if at the end of a year of unremitting labor he finds he has a few days' vacation, his eager curiosity whirls him over the vast extent of the United States, and he will travel fifteen hundred miles in a few days, to shake off his happiness.*

Tocqueville described a nation inhabited by hyperthymics.

INVENTORS, ENTREPRENEURS, AND NOBEL PRIZE WINNERS

As a nation of immigrants, the United States has racked up spectacular dopaminergic achievements. According to a research brief published by the Institute for Immigration Research at George Mason University, between 1901 and 2013 the United States received 42 percent of all

Nobel Prizes awarded, the highest of any country in the world. Moreover, a disproportionate number of American Nobel laureates have been immigrants. The top three countries they came from were Canada (13%), Germany (11%), and the United Kingdom (11%).

The United States continues to attract immigrants from all over the world, and the immigrant population continues to include a high proportion of extraordinary individuals. Some of the most important companies of the new economy were founded by immigrants, including Google, Intel, PayPal, eBay, and Snapchat. As of 2005, 52 percent of Silicon Valley start-ups had been founded by immigrant entrepreneurs, a remarkable figure in light of the fact that immigrants make up only 13 percent of the U.S. population. The country that provides America with the greatest number of technology entrepreneurs is India.

In the book *Exceptional People: How Migration Shaped Our World and Will Define Our Future*, the authors report that in 2006, foreign nationals living in the United States were listed as inventors or co-inventors on 40 percent of all international patent applications filed by the U.S. government. Immigrants also file the majority of patents by leading technology companies: 60 percent of the total at Cisco, 64 percent at General Electric, 65 percent at Merck, and 72 percent at Qualcomm.

Immigrants don't just launch technology companies. From nail salons, restaurants, and dry cleaners to the fastest-growing companies in America, immigrants start a quarter of all new businesses in the United States—about twice as many per capita as other Americans. And looking at entrepreneurship broadly, we can come full circle and find a direct link to dopamine.

A group of researchers led by Nicos Nicolaou of the Entrepreneurship & Innovation Enterprise Research Centre at the Warwick Business School recruited 1,335 people in the United Kingdom and asked them to fill out a questionnaire on entrepreneurship and to provide a blood sample for DNA extraction. The average age of the volunteers was 55 years and 83 percent were women. Nicolaou found a dopamine gene that came in two forms (alleles), identical except for one single building block. That variation in the building block (called a nucleic acid) made one form of the gene more active than the other. People who had the more active form were almost twice as likely to have started a new business compared to those who had the less active form.

It's worth noting that it's not just the United States that has been shaped by dopaminergic immigrants. The Global Entrepreneurship Monitor, an ongoing project sponsored by Babson College and the London School of Economics, found that the four nations with the highest per capita creation of new companies are the United States, Canada, Israel, and Australia—three of which are among the top nine countries with the highest immigrant populations in the world, and one of which, Israel, is less than three generations from its founding as an immigrant state.

There are a limited number of highly dopaminergic people in the world, so one country's gain is another country's loss. Many American immigrants came from Europe, a migration that boosted the dopaminergic gene pool in the United States, leaving Europe with a residual population more likely to take an H&N approach to life.[1]

The Pew Research Center conducted a survey to learn more about the differences between Americans and Europeans, and published their findings in a report titled *The American–Western European Values Gap*. Although values are influenced by many factors other than genetics, some of the questions they asked were closely related to the dopaminergic personality. For example, they asked, "Is success in life determined by forces outside our control?" In Germany, 72 percent said yes. In France it was 57 percent, and in Britain, 41 percent. However, only a little more than a third of U.S. respondents said outside forces were in control, while the majority took a more dopaminergic outlook.

The dopaminergic difference shows up in other questions, too. Americans were more likely to approve of the use of military force—the

---

1   In chapter five, we discussed ways in which American liberals, representing the party of change, tend to be more dopaminergic than conservatives, who are more likely to support maintenance of the status quo. In Europe it's reversed. Liberal governments generally represent the status quo, while the right-wing parties advocate for radical change.

literal imposition of change—to achieve national objectives. They were less likely to say it was necessary to get permission from the U.N. They also placed a greater value on religion in their lives, with 50 percent saying it was very important. Less than half that said so in Europe: 22 percent in Spain, 21 percent in Germany, 17 percent in Britain, and 13 percent in France.

The United States and other immigrant societies may have the most dopaminergic genes, but a dopaminergic approach to life has become an integral part of modern culture, whether one's genes support it or not. The world is now characterized by a never-ending flow of information, new products, advertising, and the perceived need for more. Dopamine is now associated with the most essential part of our being. Dopamine has taken over our souls.

I, DOPAMINE

Dopamine-producing cells make up 0.0005 percent of the brain. That's a tiny fraction of the cells we use to navigate our world. And yet, when we think about who we are in the deepest sense, we think about that tiny cluster of cells. We identify with our dopamine. In our minds, we *are* dopamine.

Ask a philosopher what is the essence of humanity, and it wouldn't be surprising if he said it was free will. The essence of humanity is our ability to move beyond instinct, to go beyond automatic reactions to our environment. It's the ability to weigh options, to consider higher concepts such as values and principles, and then to make a deliberate choice about how to maximize what we believe is good—whether it's love, money, or the ennobling of the soul. That's dopamine.

The academic might say that her essence is the ability to comprehend the world. It's her ability to rise above the flow of information from the physical senses to understand the *meaning* of what she perceives. She evaluates, judges, and makes predictions. She understands. That's dopamine.

The hedonist believes that his deepest self is the part of him that experiences pleasure. Whether it's wine, women, or song, his purpose in life is to maximize the rewards he gets when he pursues *more*. That's dopamine.

The artist says that the essence of her humanity is her ability to create. It's her godlike power to call into existence representations of truth and beauty that never existed before. The springs from which that creation flows are her being. That's dopamine.

Finally, the spiritual person might say that transcendence is the root of humanity. It's the thing that rises above physical reality—the most essential part of who we are is our immortal souls that exist beyond space and time. Because we cannot see, hear, smell, taste, or touch our souls, we encounter them only in our imagination. That's dopamine.

HOW TO SCRATCH YOUR HEAD

And yet, more than 99.999 percent of the brain is made up of non-dopamine-producing cells. Many of them take care of functions that are outside of our awareness, such as breathing, keeping our hormonal systems in balance, and coordinating muscles that allow us to carry out seemingly simple motions. Think about scratching your head. It starts out with your dopamine circuits deciding it's a good idea. They decide that scratching your head is the best path to an itch-free future. Dopamine cells give the signal to do it, but that's where dopamine—and conscious involvement—come to an end.

Dopamine is the conductor, not the orchestra.

In some ways the dopaminergic command, *do it*, is the easiest part. What comes next is so complicated, it's hard to even imagine how we get it done.

Lifting your arm to scratch your head requires the coordination of dozens of muscles in your fingers, wrist, arm, shoulder, back, neck, and abdomen. If you're standing when you do it, the coordination requirements go all the way down your legs. Moving your arm upward changes your center of gravity, so it requires balance adjustments. It's complicated. Each joint in your body has opposing muscles (similar to the opposing circuits in the brain) so that the joint can be controlled with a high degree of precision. The muscles on one side of the joint need to contract with a specific and constantly changing strength, while the opposing ones have to relax in a constantly changing manner. Muscles are made of individual fibers. There are a quarter million of them in your biceps alone. The strength of contraction depends on what percent of these fibers are being activated, so each fiber needs to be controlled separately. To scratch your head, your brain must control millions of muscle fibers throughout your body. It must make sure they are all properly coordinated with each other and dynamically modify the relative strength of contraction over the course of the movement. That requires a lot of brainpower. Probably more than you knew you had. It's not dopamine, but it's still you.

Much of what we do throughout the day is automatic. We walk out the door and go to work with little intentional thought. We drive cars, feed ourselves, laugh, smile, frown, slouch, and do thousands of other things without having to think about them. We do so much that bypasses the part of the brain that weighs options and makes choices, that an argument could be made that those non-conscious actions—non-dopaminergic activities—represent who we really are.

SHE'S NOT HERSELF TODAY

The people we know and love all have special characteristics that define who they are. Some of those characteristics arise from dopamine activity. We might say, "He's always there when you need him." But often a person's unconscious, non-dopaminergic actions are even more precious to us. We might say things like, "She's always happy. No matter how bad I feel, she can cheer me up." "I love the way he smiles." "She has the most bizarre sense of humor." "There's something about the way he walks that is so him."

The way those individual muscle fibers contract to get our arm up to our head when we scratch might not seem particularly relevant to the essence of our being, but our friends might disagree. Each one of us has a unique way of moving. We're usually unaware of these habits, but other people see them. Often we recognize our friends from a distance based on how they move, even when we can't see their faces. The way we move is part of what defines us.

What do we mean when we say, "She's not herself today"? The person might be sick; she might be feeling weighed down by disappointment; she may be tired because she didn't sleep last night. Whatever it is, it rarely means that our friend is choosing to act like a different person. It generally means that aspects of her behavior that are outside her conscious control are different. And it's those aspects that we refer to when we think of "herself"—the essence of who she is. We may believe our souls reside in our dopamine circuits, but our friends don't believe that.

What else do we neglect when we identify our core being with our dopamine circuits? We neglect emotion, empathy, the joy of being with people we care about. If we ignore our emotions, lose touch with them, they become less sophisticated over time, and may devolve into anger, greed, and resentment. If we neglect empathy, we lose the ability to make others feel happy. And if we neglect affiliative relationships, we will most likely lose the ability to be happy ourselves—and probably die early. A Harvard study that's been going on for seventy-four years has found that social isolation (even in the absence of feelings of loneliness) is associated with a 50 to 90 percent higher risk of early death. That's about the same as smoking, and higher than obesity or lack of exercise. Our brain needs affiliative relationships just to stay alive.

We also lose the pleasure of the sensory world around us. Instead of enjoying the beauty of a flower, we imagine only how it would look in a vase on our kitchen table. Instead of smelling the morning air and looking at the sky, we consult the weather app on our smartphone, neck bent, oblivious to the world around us.

Identifying ourselves with our dopamine circuits traps us in a world of speculation and possibility. The concrete world of here and now is disdained, ignored, or even feared, because we can't control it. We can only control the future, and giving up control is not something dopaminergic creatures like to do. But none of it is real. Even a future one second away is unreal. It is only the stark facts of the present that are real, facts that must be accepted exactly as they are, facts that cannot be modified by a hair's breadth to suit our needs. This is the world of reality. The future, where dopaminergic creatures live their lives, is a world of phantoms.

Our worlds of fantasy can become narcissistic havens where we are powerful, beautiful, and adored. Or perhaps they're worlds where we are in total control of our environment the way a digital artist controls every pixel on his screen. As we glide through the real world, half blind, caring only about things we can put to use, we trade the deep oceans of reality for the shallow rapids of our never-ending desires. And in the end, it might annihilate us.

WILL DOPAMINE DESTROY THE HUMAN RACE?

When the human race lived in scarcity and on the brink of extinction, the drive for *more* kept us alive. Dopamine was the engine of progress. It helped lift our evolutionary ancestors out of subsistence living. By giving us the ability to create tools, invent abstract sciences, and plan far into the future, it made us the dominant species on the planet. But in an environment of plenty in which we have mastered our world and developed sophisticated technology—in a time when *more* is no longer a matter of survival—dopamine continues to drive us forward, perhaps to our own destruction.

As a species we have become far more powerful than we were when our brains first developed. Technology develops fast while evolution is slow. Our brains evolved at a time when survival was in doubt. That's less of a problem in the modern world, but we're stuck with our ancient brains.

It's possible that we won't last beyond another half-dozen genera-tions. We've simply become too good at gratifying our dopaminergic desires: not all forms of *more* and *new* and *novel* are good for an individual, and the same is true for a species. Dopamine doesn't stop. It drives us ever onward into the abyss. In the following sections we're going to look at worst-case scenarios. It may be that our dopaminergic-driven ingenuity will help us find a safe way through the reefs and shoals of humanity's ever-accelerating progress. Then again, maybe not. For example:

PRESS THE BUTTON

Nuclear armageddon is the most obvious way in which dopamine can destroy humanity. Highly dopaminergic scientists have built doomsday weapons for highly dopaminergic rulers. Scientists can't stop themselves from making their weapons ever more deadly, and dictators can't help themselves from lusting after power. Over time, more and more countries are acquiring nuclear capabilities, and someday someone's dopamine circuits might come to the conclusion that the best way to maximize future resources is to press the button. We all hope—and many believe—that before we destroy ourselves, humanity will find a way to move beyond our primitive drive for conquest, possibly through organizations of international cooperation such as the United Nations.

But if that happens, it's going to take something very powerful to bring it about. It's awfully hard to rewire our brains.

FINISH OFF THE PLANET

Another obvious doomsday scenario involves dopamine driving us on to greater and greater consumption until we destroy the planet. Climate change accelerated by industrial activity is a major focus of countries around the world that fear devastating consequences, including drought, floods, and violent competition for diminishing resources. More than half of greenhouse gases are generated by burning fossil fuel to make cement, steel, plastics, and chemicals. As more countries are lifted out of poverty, the demand for these materials increases. Everybody wants *more*—and for a significant plurality of nations, *more* isn't the pursuit of luxury. It's the climb out of crushing poverty.

The Intergovernmental Panel on Climate Change, which provides scientific assessments for the United Nations Climate Conference, asserts that any response must include fundamental social change. Global economic growth will have to be slowed down. People will need to use less heat, less air conditioning, less hot water. They will have to drive less, fly less, and consume less. In other words, behavior driven by dopamine will need to be drastically suppressed and the era of better, faster, cheaper, and more will have to end.

This has never happened in the history of humanity—at least not by our choice. Only breakthrough technologies will allow us to continue our current rate of rising consumption, while reducing the production of greenhouse gases.

LET'S ALL WELCOME OUR NEW SILICON OVERLORDS

Computers that are smarter than people will fundamentally change the world. Every year we make faster and more powerful computers thanks to our dopamine-driven ability to use abstract concepts to create new technology. Once computers become smart enough to build—and improve—themselves, their progress will accelerate dramatically. At that point no one knows what will happen. It's possible it will occur sooner than we think. Ray Kurzweil, the world's leading futurologist, believes that we will have superintelligent computers as early as the year 2029.

Computers that are programmed using traditional techniques are completely predictable. They follow a clear set of instructions to get from the beginning of a calculation to the end. Newer developments in artificial intelligence, however, create unpredictable results. Instead of the programmer determining how the computer works, the computer modifies itself based on how successful it is in reaching its goal. It optimizes its programming to solve problems. It's called *evolutionary computing*. Circuits that lead to success are strengthened, and those that lead to failure are weakened. As the process goes on, the computer gets better and better at its assigned task—recognizing faces, for example. But no one can tell how it does it. As adjustments are made over time, the circuits become too complex to understand.

As a result, no one knows precisely what a superintelligent computer might do. An artificial intelligence that programs its own circuits might one day come to the conclusion that eliminating the human race is the best way to accomplish its goal. Scientists can try to program in safeguards, but since the program evolves outside the programmers' control, it's impossible to know what kinds of safeguards will be robust enough to survive the process of "optimization." One option is to simply stop making computers with artificial intelligence. However, that would diminish our ability to pursue *more*, so we can rule that out. Dopamine will drive the science forward whether it's good for us or not. We may get lucky, though. We may discover a way to ensure that artificial intelligences act in ethical ways. Many experts in the field believe that should be a top priority for computer scientists.

EVERYTHING. ALL THE TIME.

Dopamine-driven technological advances make it ever easier for us to gratify our needs and desires. Grocery store shelves are packed with constantly changing "new and improved" products. Planes, trains, and automobiles take us wherever we want to go, cheaper and faster than ever before. The internet provides us with virtually unlimited entertainment options, and so much cool stuff is brought to market each year that we need crowds of journalists to keep us up to date on new ways to spend our money.

Dopamine drives our lives faster and faster. It takes more education to keep up. A graduate degree is as necessary today as a college education was a generation ago. We work longer hours. There are more memos to read, reports to write, and emails to be answered. It never stops. We are expected to be available at all times of the day and night. When someone at work wants us, we must respond immediately. Advertisements show a smiling man responding to texts on the beach, or a woman by the hotel pool, checking her cell phone screen to tap into a video feed of her empty house. *What a relief*. Nothing happened since the last time she checked, 15 minutes ago. She's got everything under control.

With so many ways to have fun, so many years to devote to education, and so much time to spend working, something has to give, and that something is family. According to the U.S. Census Bureau, between 1976 and 2012 the number of childless women in America approximately doubled. *The New York Times* reports that 2015 brought the first NotMom Summit, a global gathering of women without children by choice or circumstance.

In developed countries, people have pretty much lost interest in having children. Raising kids costs a lot of money. According to the U.S. Department of Agriculture it costs $245,000 to raise a child to the age of eighteen. Four years of college tuition plus room and board costs another $160,000, and after college there's graduate school, or maybe the kids will move back home. Add it all together and you might be able to buy a vacation home or travel overseas every year, not to mention restaurants, the theater, and designer clothes. As one newlywed who planned to have no children succinctly put it, "More money for us."

Future-focused dopamine no longer drives couples to have children because people who live in developed countries don't depend on their children to support them in their old age. Government-funded retirement plans take care of that. That frees up dopamine to move on to other things like TVs, cars, and remodeled kitchens.

The end result is demographic collapse. About half the world lives in a country with below replacement fertility. Replacement fertility is the number of children each couple must have to prevent a decline in the population. In developed countries the number is 2.1 per woman in order to replace the parents, and a bit more to account for early deaths. In some developing countries replacement fertility is as high as 3.4 because of high rates of infant mortality. The worldwide average is 2.3.

All European countries as well as Australia, Canada, Japan, South Korea, and New Zealand have transitioned to below-replacement fertility rates. The United States has enjoyed a more stable rate, largely because of the influx of immigrants from developing countries who haven't yet lost the habit of continuing the survival of the human race.

But even in developing countries birth rates are falling. Brazil, China, Costa Rica, Iran, Lebanon, Singapore, Thailand, Tunisia, and Vietnam have all transitioned to below-replacement fertility rates.

Governments are doing what they can to prevent their countries from becoming ghost towns. During the Syrian refugee crisis, Germany famously opened its borders to all comers. Denmark responded to the baby crisis by creating commercials showing a sultry model wearing a black negligee, encouraging viewers to "Do it for Denmark." Singapore, which has a birthrate of only 0.78, made a deal with Mentos ("The Freshmaker") to promote "National Night" in which couples were told to let their "patriotism explode." In South Korea couples earn cash and prizes for having more than one child, and in Russia they get a chance to win a refrigerator.

DO NOTHING, EXPERIENCE EVERYTHING

Finally, the decline if not the end of the human race may be accelerated by virtual reality (VR). VR already creates compelling experiences in which the participant is transported to beautiful, exciting locations to become the hero of the universe—instantly.

VR produces images and sound, with other sensory modalities coming online soon. For instance, researchers in Singapore have developed what they call a "digital taste simulator." It's a device with electrodes that deliver current and heat to the tongue. By stimulating the tongue with varying amounts of electricity and heat, it's possible to trick it into experiencing salty, sour, and bitter flavors. Other groups have managed to simulate sweet as well. Once scientists master all the basic flavors, they'll be able to combine them in different proportions to allow the tongue to experience the sensation of tasting almost any food imaginable. Since what we perceive as taste is, in large part, smell, there's also a device that features an aromatic diffuser that simulates smells. It comes with what the inventors call a "bone conduction transducer." They say that it "mimics the chewing sounds that are transmitted from the diner's mouth to ear drums via soft tissues and bones."

Touch is the final frontier, since that will allow VR makers to simulate sex, and pornography is the universal driver of new media adoption, such as VCRs, DVDs, and high-speed internet. Why bother having sex with a needy, repetitive, imperfect partner when an ever-changing fantasy can be had instead? Pornography is about to become a lot more addictive by entering the realm of touch. Devices have recently come to market that deliver genital stimulation synchronized with pornographic VR—essentially sex toys manipulated by a computer. There's a lot of money at stake. In 2016 the market for sex toys was $15 billion, with projections that it will surpass $50 billion by 2020.

Soon we'll be able to teach the computer what we like by rating the experiences it generates in the same way we rate music and books. The computer will become so adept at fulfilling our desires that no human will be able to compete. The next step will be bodysuits that will allow us to experience virtual sex with all our senses, without the inconvenience of reproduction. People are already choosing to have fewer children. When current trends meet the allure of VR, the future of the human race will be very much in doubt.

With VR, the human race may go willingly into the dark night. Our dopamine circuits will tell us it's the best thing ever.

There's only one thing that will save us: the ability to achieve a better balance, to overcome our obsession with *more*, appreciate the unlimited complexity of reality, and learn to enjoy the things we have.

FURTHER READING

Huff, C. D., Xing, J., Rogers, A. R., Witherspoon, D., & Jorde, L. B. (2010). Mobile elements reveal small population size in the ancient ancestors of *Homo sapiens*. *Proceedings of the National Academy of Sciences, 107*(5), 2147–2152.

Chen, C., Burton, M., Greenberger, E., & Dmitrieva, J. (1999). Population migration and the variation of dopamine D4 receptor (DRD4) allele frequencies around the globe. *Evolution and Human Behavior, 20*(5), 309–324.

Merikangas, K. R., Jin, R., He, J. P., Kessler, R. C., Lee, S., Sampson, N. A., . . . Ladea, M. (2011). Prevalence and correlates of bipolar spectrum disorder in the World Mental Health Survey Initiative. *Archives of General Psychiatry, 68*(3), 241–251.

Keller, M. C., & Visscher, P. M. (2015). Genetic variation links creativity to psychiatric disorders. *Nature Neuroscience, 18*(7), 928.

Smith, D. J., Anderson, J., Zammit, S., Meyer, T. D., Pell, J. P., & Mackay, D. (2015). Childhood IQ and risk of bipolar disorder in adulthood: Prospective birth cohort study. *British Journal of Psychiatry Open, 1*(1), 74–80.

Bellivier, F., Etain, B., Malafosse, A., Henry, C., Kahn, J. P., Elgrabli-Wajsbrot, O., . . . Grochocinski, V. (2014). Age at onset in bipolar I affective disorder in the USA and Europe. *World Journal of Biological Psychiatry, 15*(5), 369–376.

Birmaher, B., Axelson, D., Monk, K., Kalas, C., Goldstein, B., Hickey, M. B., . . . Kupfer, D. (2009). Lifetime psychiatric disorders in school-aged offspring of parents with bipolar disorder: The Pittsburgh Bipolar Offspring study. *Archives of General Psychiatry, 66*(3), 287–296.

Angst, J. (2007). The bipolar spectrum. *The British Journal of Psychiatry, 190*(3), 189–191.

Akiskal, H. S., Khani, M. K., & Scott-Strauss, A. (1979). Cyclothymic temperamental disorders. *Psychiatric Clinics of North America, 2*(3), 527–554.

Boucher, J. (2013). *The Nobel Prize: Excellence among immigrants*. George Mason University Institute for Immigration Research.

Wadhwa, V., Saxenian, A., & Siciliano, F. D. (2012, October). *Then and now: America's new immigrant entrepreneurs, part VII*. Kansas City, MO: Ewing Marion Kauffman Foundation.

Bluestein, A. (2015, February). The most entrepreneurial group in America wasn't born in America. Retrieved from http://www.inc.com/magazine/201502/adam-bluestein/the-most-entrepreneurial-group-in-america-wasnt-born-in-america.html

Nicolaou, N., Shane, S., Adi, G., Mangino, M., & Harris, J. (2011). A polymorphism associated with entrepreneurship: Evidence from dopamine receptor candidate genes. *Small Business Economics, 36*(2), 151–155.

Kohut, A., Wike, R., Horowitz, J. M., Poushter, J., Barker, C., Bell, J., & Gross, E. M. (2011). *The American-Western European values gap.* Washington, DC: Pew Research Center.

Intergovernmental Panel on Climate Change. (2014). IPCC, 2014: Summary for policymakers. In *Climate change 2014: Mitigation of climate change* (Contribution of Working Group III to the Fifth Assessment Report of the Intergovernmental Panel on Climate Change). New York, NY: Cambridge University Press.

Kurzweil, R. (2005). *The singularity is near: When humans transcend biology.* New York: Penguin.

Eiben, A. E., & Smith, J. E. (2003). *Introduction to evolutionary computing* (Vol. 53). Heidelberg: Springer.

Lino, M. (2014). Expenditures on children by families, 2013. Washington, DC: U.S. Department of Agriculture.

Roser, M. (2017, December 2). Fertility rate. *Our World In Data.* Retrieved from https://ourworldindata.org/fertility/

McRobbie, L. R. (2016, May 11). 6 Creative ways countries have tried to up their birth rates. Retrieved from http://mentalfloss.com/article/33485/6-creative-ways-countries-have-tried-their-birth-rates

Ranasinghe, N., Nakatsu, R., Nii, H., & Gopalakrishnakone, P. (2012, June). Tongue mounted interface for digitally actuating the sense of taste. In *2012 16th International Symposium on Wearable Computers* (pp. 80–87). Piscataway, NJ: IEEE.

Project Nourished—A gastronomical virtual reality experience. (2017). Retrieved from http://www.projectnourished.com

Burns, J. (2016, July 15). How the "niche" sex toy market grew into an unstoppable $15B industry. Retrieved from http://www.forbes.com/sites/janetwburns/2016/07/15/adult-expo-founders-talk-15b-sex-toy-industry-after-20-years-in-the-fray/#58ce740538a1

Do you wish to be great? Then begin by being. Do you desire to construct a vast and lofty fabric? . . . The higher your structure is to be, the deeper must be its foundation.
—Saint Augustine

I arise in the morning torn between a desire to improve the world and a desire to enjoy the world. This makes it hard to plan the day.
—E. B. White

Chapter 7

HARMONY

Putting it all together.

In which dopamine and H&Ns find balance.

THE DELICATE BALANCE BETWEEN DOPAMINE AND THE H&NS

A middle-aged man went to see a specialist to have his depression treated. In addition to feeling sad and hopeless, he had an unhealthy obsession with the future. He ruminated over everything that might go wrong, constantly fearful of some unknown catastrophe. His psychic energy was drained by the worry, and he became emotionally brittle. He blew up at the slightest provocation. He was unable to take the train to work because it was intolerable to be jostled or even touched by other riders. There were nights when his wife woke up at 3 AM to find him in tears. He said, "When you get a flat tire, an ordinary person calls the AAA. I call the suicide hotline."

He was given the standard treatment for depression, an antidepressant that changes the way the brain uses the H&N neurotransmitter serotonin, and he had an excellent response. Over the course of about a month his mood gradually improved until he was once again bright and cheerful. He became more resilient and was able to enjoy the good things in his life. It was a relief to his wife, as well. He thought it would be interesting to try a higher dose of the medication, just to see what would happen, and his doctor agreed. "It felt great," he said at his next visit. "I was so happy, there was nothing I needed to do. There was no reason to get out of bed in the morning." He and his doctor decided to reduce the dose to its previous level, and his emotional balance returned.

The dramatic reaction this patient had to a serotonergic antidepressant occurs in only a few people who have just the right combination of genes and environment. But it's a good illustration of how a person can be disabled by both an excessive focus on the future and an excessive enjoyment of the present.

Dopamine and the H&N neurotransmitters evolved to work together. They often act in opposition to one another, but that helps maintain stability among constantly firing brain cells. In many instances, though, dopamine and H&N get thrown out of balance, especially on the dopaminergic side. The modern world drives us to be all dopamine, all the time. Too much dopamine can lead to productive misery, while too much H&N can lead to happy indolence: the workaholic executive versus the pot-smoking basement dweller. Neither one is living a truly happy life or growing as a person. To live a good life, we need to bring them back into balance.

We instinctively know that neither extreme is healthy, and that may be one of the reasons we like stories about people who start out with too much of one or the other and in the end find balance. The movie *Avatar* provides an example of someone who starts out with too much dopamine. A former Marine named Jake is hired to work for the security arm of a mining company. The company is intent on exploiting the natural resources of a moon called Pandora, which is covered by undisturbed forests and populated by the Na'vi, a race of humanoids who live in harmony with nature. The Na'vi worship a mother goddess called Eywa. It's a classic example of dopamine versus H&N.

To maximize the resources they can dig up, the mining company plans to destroy the sacred Tree of Souls, which is in their way. Appalled at the plan, Jake rejects his dopaminergic background, joins the H&N Na'vi, and develops close, affiliative relationships with members of the tribe. Combining his dopaminergic skills with his newly acquired ability to work together with the Na'vi, he organizes them and leads them to victory against the security forces of the mining company. In the end, with the help of the Tree of Souls, Jake becomes one of the Na'vi and achieves balance.

The classic 1980s movie *Trading Places* takes us to a place of balance from the opposite direction. Billy Ray Valentine is an irresponsible homeless man. He's lazy, indulgent, and doesn't give any thought for the future. He becomes the subject of an experiment in which his life is swapped with that of a successful commodities trader, who is his mirror image. As Billy Ray accumulates wealth, he rejects his former carefree lifestyle and becomes responsible. In one scene he invites a group of old friends to a party at his mansion and is uncharacteristically upset when they vomit on his Persian rug. In the end he participates in an elaborately planned scheme that makes him rich, and returns him to a life of leisure, but with a new set of capabilities.

How can the ordinary person find balance? It's unlikely that any of us will forsake the modern world to live with a clan of tree-worshipers. We have to find balance in other ways. Dopamine alone will never satisfy us. It can't provide satisfaction any more than a hammer can turn a screw. But it's constantly promising us that satisfaction is right around the corner: one more donut, one more promotion, one more conquest. How do we get off the treadmill? It's not easy, but there are ways.

MASTERY: THE PLEASURE OF BEING GOOD AT SOMETHING

Mastery is the ability to extract the maximum reward from a particular set of circumstances. One might achieve mastery over *Pac-Man*, racquetball, French cooking, or debugging a complicated computer program. From dopamine's point of view mastery is a good thing—something to be desired and pursued. But it's different from other good things. It's not simply finding food, or a new partner, or beating the competition. It's bigger and broader than that. It's reward extraction success: dopamine achieving dopamine's goal. When mastery is achieved, dopamine has reached the pinnacle of its aspiration—squeezing every last drop out of an available resource. This is what it's all about. This is the moment to savor—now, in the present. Mastery is the point at which dopamine bows to H&N. Having done all it can do, dopamine pauses, and allows H&N to have its way with our happiness circuits. Even if it's only for a short time, dopamine doesn't fight the feeling of contentment. It approves. The best basking is basking in a job well done.

Mastery also creates a feeling of what psychologists call an *internal locus of control*. This phrase refers to the tendency to view one's choices and experiences as being under one's own control as opposed to being determined by fate, luck, or other people. It's a good feeling. Most people don't like being at the mercy of forces beyond their control. Pilots say that when they're flying in bad weather, it's less stressful to be at the controls than to sit in the cabin. It's the same with driving in a snowstorm. Most people would rather be in the driver's seat than in the passenger seat. In addition to making people feel good, an internal locus of control also makes people more effective. People with a strong sense of internal locus of control are more likely to achieve academic success and get high-paying jobs.

Those who have an *external locus of control*, by contrast, take a more passive view of life. Some are happy, relaxed, and easygoing, but at the same time they often blame others for their failures and may not put forth their best effort on a consistent basis. Doctors often become frustrated with this kind of person. They tend to ignore medical advice, and they aren't easily persuaded to accept responsibility for their health by taking their medication every day and making healthy lifestyle choices.

The development of an internal locus of control, as well as contentment (if only for a little while), are among the many benefits of achieving mastery over an activity. But it takes an enormous amount of time and effort as well as constant mental stretching. Mastering a skill requires a student to constantly move outside her comfort zone. As soon as a piano player gets good at an easy song, she has to start on a harder one. It's a tough slog, but it can also be a great joy. Those who don't give up generally feel it was worth it. It can result in a feeling that they have found their passion, something so engrossing they become completely immersed in it.

THE REWARDS OF REALITY

What do you think about when you brush your teeth? Probably not brushing your teeth. You're more likely to be thinking about things you have to do later in the day, later in the week, or some other time in the future. Why? Maybe it's a habit. Maybe it's anxiety. Maybe you're afraid that if you don't think about the future you will miss something. But you probably won't. And by not thinking about what you're doing, you will definitely miss something, maybe even something you never noticed before, something unexpected.

What dopamine loves more than anything else is reward prediction error, which, as we have discussed, is the discovery that something is better than we had anticipated it would be. Paradoxically, dopamine does everything in its power to avoid such incorrect forecasts. Reward prediction error feels great because your dopamine circuits get excited over the fact that there is something new and unexpected to make your life better. But being surprised by an unexpected new resource means the resource isn't being fully exploited. So dopamine makes sure the surprise that felt so good will never be a surprise again. Dopamine extinguishes its own pleasure. It's frustrating, but it's the best way to keep us alive. What can we do to keep the surprises coming?

Reality is the richest source of the unexpected. Fantasies that we conjure in our minds are predictable. We go over the same material again and again. Once in a while we'll be struck by an original idea, but it's rare, and it usually happens when we're paying attention to something else—not when we're trying to strong-arm our creativity into action.

Paying attention to reality, to what you are actually doing in the moment, maximizes the flow of information into your brain. It maximizes dopamine's ability to make new plans, because to build models that will accurately predict the future, dopamine needs data, and data flows from the senses. That's dopamine and H&N working together.

When something interesting activates the dopamine system, we snap to attention. If we are able to activate our H&N system by shifting our focus outward, the increased level of attention makes the sensory experience more intense. Imagine walking down a street in a foreign country. Everything is more exciting, even looking at ordinary buildings, trees, and shops. Because we are in a novel situation, sensory inputs are more vivid. That's a large part of the joy of travel. It works in the opposite direction, too. Experiencing H&N sensory stimulation, especially within a complex environment (sometimes called an *enriched environment*), makes the dopaminergic cognitive facilities in our brains work better. The most complex environments, those that are most enriched, are usually natural ones.

GO AHEAD AND TAKE A MICROBREAK . . .

Nature is complex. It's made up of systems with many interacting parts. Unexpected patterns emerge as a result of a large number of elements influencing one another. There's a virtually limitless amount of detail to explore. We also perceive it as beautiful, inspiring, sometimes calming, and other times energizing. Dr. Kate Lee and a team of researchers at the University of Melbourne, Australia, tested the cognitive effects of a mere 40 seconds of exposure to nature in the form of a picture of a city building with grass and flowers covering the roof. They compared it to the effects of a picture of a similar building covered with concrete.

To measure the impact of these different scenes, the researchers asked a group of students to perform a concentration task. Random numbers were flashed on a screen, and the students had to press a button as soon as they saw the number. But they had to hold back when the number was 3. They had less than a second to react, and they had to do it 225 times in a row. It's a hard task that requires a great deal of concentration and motivation to get it right. The researchers asked the students to do the task twice, with a 40-second "microbreak" in between.

Students who looked at the picture of flowers and grass between the first and second trials made fewer errors than those who looked at the concrete roof. The researchers speculated that the most likely explanation for the difference was that the natural scene stimulated both "sub-cortical arousal" (desire dopamine) and "cortical attention control" (control dopamine). A reporter from the *Washington Post* who commented on the study noted that "urban rooftops covered with grasses, plants and other types of greenery are becoming increasingly popular around the world . . . [Facebook] recently installed a massive 9-acre green roof at its office in Menlo Park, California." That approach to architecture, using H&N stimulation to activate dopamine, is not only good for the soul—it may also be good for the bottom line.

. . . BUT DON'T TRY TO MULTITASK

Almost any experience is improved by paying full attention to it.
—Kelly McGonigal, Lecturer in Management,
Stanford School of Business

In spite of what technology addicts may believe, multitasking, or paying attention to more than one thing at a time, is impossible. When you attempt to do more than one thing, such as talking on the telephone while reading an email, you shift your attention between the tasks, and end up compromising on both. Sometimes you pause while reading the email to listen to the person on the phone; other times you stop listening as you focus on the email. The person you're talking to can tell. You're obviously not giving him your full attention, and you miss important details. Instead of increasing your efficiency, "multitasking" decreases it.

Aza Raskin, an expert on user experience and the lead designer for the Firefox 4 web browser, gives an example. Spell aloud, letter by letter, "Jewelry is shiny" while at the same time printing your name. How long does it take? Now spell aloud, letter by letter, "Jewelry is shiny" and then, after you are done with that, write your name. How long did that take? Probably about half as much time as "multitasking" did.

You also make more mistakes when you try to multitask. Interruptions of only a few seconds, the amount of time it takes to switch to your email program and back, can double the number of errors you make on a task that requires concentration. It's not just the distraction that causes the mistakes; switching back and forth consumes mental energy, and fatigue makes it harder to concentrate. Still, people do it, especially people who work with computers.

A study from the University of California, Irvine, in collaboration with Microsoft and the Massachusetts Institute of Technology tracked the work habits of people who spent most of their day online. The average amount of time they spent on one task before switching to another was only 47 seconds. Over the course of the day they switched between tasks more than four hundred times. Those who spent less time before jumping to something else experienced higher levels of stress and got less work done—if for no other reason than that they repeated the "switch tasks" maneuver four hundred times instead of only once after each task was completed. In addition to decreasing productivity, high levels of stress also cause fatigue and burnout.

THE HIGH COST OF LIVING IN THE FUTURE

Living our lives in the abstract, unreal, dopaminergic world of future possibilities comes at a cost, and that cost is happiness. Researchers from Harvard University discovered this by developing a smartphone app that prompted volunteers to provide real-time reports of their thoughts, feelings, and actions as they went about their daily activities. The goal of the study was to learn more about the relationship between a wandering mind and happiness. Over five thousand people from eighty-three countries volunteered to be in the study.

The app contacted the participants at random times to request data. It asked the volunteers, "How are you feeling right now?" "What are you doing right now?" and "Are you thinking about something other than what you're currently doing?" People answered *yes* to the last question about half the time, no matter what they were doing. All activities produced the same amount of mind wandering except sex, which was very good at keeping people's attention. In every other situation, though, thinking about other things happened so frequently that the researchers concluded that a wandering mind, what scientists call *stimulus-independent thought*, was the brain's default mode.

When they looked at happiness, they found that people were less happy when their mind was wandering, and once again, it didn't matter what the activity was. Whether they were eating, working, watching TV, or socializing, they were happier if they were paying attention to what they were doing. They researchers concluded that "a human mind is a wandering mind, and a wandering mind is an unhappy mind."

But what if you don't care about happiness? What if you're so dopaminergic that the only thing you care about is achievement? It doesn't matter, because no matter how brilliant, original, or creative you are, your dopamine circuits aren't going to achieve much without the raw material provided by the H&N senses.

Michelangelo's *Pietà*, depicting the Virgin Mary cradling her dead son, powerfully communicates the abstract ideas of grief and acceptance. But it took a block of marble to realize the artist's conception. The sad beauty of Mary is an idealized depiction of femininity, but Michelangelo could not have conceived this image had he not used his eyes to study real women and his emotions to feel real sorrow in the here and now.

By spending time in the present, we take in sensory information about the reality we live in, allowing the dopamine system to use that information to develop reward-maximizing plans. The impressions that we absorb have the potential to inspire a flurry of new ideas, enhancing our ability to find new solutions to the problems we face. And that's a wonderful thing. Creating something new, something that has never been conceived of before is, by definition, surprising. Because it is always new, creation is the most durable of the dopaminergic pleasures.

STIR IT UP

Creativity is an excellent way to mix together dopamine and H&N. We discussed a particular kind of creativity in chapter four, a creativity achieved by dismantling conventional models of reality. It's an extraordinary creativity in which the creator is driven to pursue his work to the exclusion of all other aspects of life, such as family and friends. Lonely and obsessed, people with breakthrough ideas are usually dissatisfied. Dopamine predominates, and H&N circuits wither. But there are more ordinary forms of creativity that anyone can practice, acts of creation that promote balance, rather than dopaminergic dominance.

Woodworking, knitting, painting, decorating, and sewing are old-fashioned activities that don't get much attention in our modern world—which is exactly the point. These activities don't require smartphone apps or high-speed internet. They require brains and hands working together to create something new. Our imagination conceives the project. We develop a plan to carry it out. Then our hands make it real.

A business executive working in financial services spent his days brooding over stock options, asset derivatives, foreign exchange rates, and other imaginary beasts. He was wealthy and miserable. His misery drove him to see a mental health specialist, and a few months later he had rediscovered his passion for painting, a hobby he had abandoned decades ago. "I can't wait to get home at the end of the day," he told his doctor. "Last night I painted for four hours, and I didn't even realize the time had gone by."

Not everyone has the time or inclination to learn how to paint, but that doesn't mean that creating beauty is out of reach. Coloring books for adults have mystified some and satisfied many. At first glance they seem silly; why do adults need coloring books? But they have the ability to relieve stress by providing an escape from the imbalanced, dopaminergic world. Coloring books for adults feature beautiful, abstract geometric patterns—dopaminergic abstractions combined with sensory experience.

Children also need to work with their hands. In 2015, *Time* magazine published an article titled "Why Schools Need to Bring Back Shop Class." Working with drills and rip saws, surrounded by the aroma of fresh sawdust, is a welcome break from the intellectual rigors of academic classes. Sanding a piece of wood until it's "as smooth as a baby's bottom," as one shop teacher put it, is a joy that few people experience these days. And the birdhouse that comes into being at the end of it all—it's a small miracle. Dwelling on it is an oasis of peace where the mind can say, *I made that.*

Many people grew up in a home where their father had a workbench in the garage. They're less common today, but fixing things is a unique pleasure. Each project is a problem that needs to be solved—a dopaminergic activity—and then the solution is made real. Sometimes solving repair problems requires creativity because the necessary tools or supplies aren't available. For example, figuring out that a nail clipper can be used as a wire cutter. Fixing things also boosts self-efficacy and increases one's sense of control: H&N delivering dopaminergic gratification.

Cooking, gardening, and playing sports are among many activities that combine intellectual stimulation with physical activity in a way that will satisfy us and make us whole. These activities can be pursued for a lifetime without becoming stale. You might get a few weeks of dopaminergic thrills by buying an expensive Swiss timepiece, but after that it's just a watch. Getting promoted to district manager makes going to work exciting at first, but eventually it becomes the same old grind. Creativity is different because it stirs together H&N with dopamine. It's like mixing little bit of carbon with iron to make steel. The result is stronger and more durable. That's what happens to dopaminergic pleasure when you add physical H&N.

But most people don't bother to engage in acts of creation, like drawing pictures, making music, or building model airplanes. There's no practical reason to do these things. They're hard, at least in the beginning, and they probably won't earn us money or prestige or guarantee us a better future. But they might make us happy.

THE POWER IS IN YOUR HANDS

In 2015, TINYpulse, a consulting firm that helps managers increase employee engagement, surveyed over 30,000 employees working for more than five hundred companies. They asked the employees about their managers, their coworkers, and professional growth. But what the survey was really about was happiness.

TINYpulse noted that no one had ever performed a survey like this. Management consultants in general didn't seem to place much value on happiness. But TINYpulse believed that happiness was essential to a company's success, so they looked at happiness in a broad range of industries, including the glamorous fields of technology, finance, and biotech. None of them came out on top. The happiest people were construction workers.

Construction workers take abstract plans and make them real. They use their minds and their hands. They also enjoy a high degree of camaraderie. When TINYpulse looked at the reasons construction workers gave for feeling happy, the most common was, "I work with great people." A construction manager said, "One thing that unites everybody at the end of the day is kicking back for a little bit with a few beers and talking stuff out—the good and the bad." Affiliative relationships in the context of the work environment played a key role: work and friendship, dopamine and H&N.

The second most important reason for happiness given by construction workers was, "I'm excited about my work and projects," a dopaminergic reason. The authors of the report also noted that the construction industry had enjoyed strong growth in the previous year, and this growth was reflected in rising salaries, another dopaminergic contribution. It takes both dopamine and H&N to attain happiness, the state of being that the philosopher Aristotle considered to be the goal of all other goals.

Our dopamine circuits are what make us human. They are what give our species its special power. We think. We plan. We imagine. We elevate our thoughts to ponder abstract concepts such as truth, justice, and beauty. Within those circuits we transcend all barriers of space and time. We thrive in the most hostile environments—even in outer space—thanks to our ability to dominate the world around us. But these same circuits can also lead us down a darker path, a path of addiction, betrayal, and misery. If we aim to be great, we will probably have to accept the fact that misery will be a part of it. It's the goad of dissatisfaction that keeps us at our work while others are enjoying the company of family and friends.

But those of us who prefer a life of happy fulfillment have a different task to accomplish: the task of finding harmony. We have to overcome the seduction of endless dopaminergic stimulation and turn our backs on our never-ending hunger for more. If we are able to intermingle dopamine with H&N, we can achieve that harmony. All dopamine all the time is not the path to the best possible future. It's sensory reality and abstract thought working together that unlocks the brain's full potential. Operating at its peak performance, it becomes capable of producing not only happiness and satisfaction, not only wealth and knowledge, but a rich mixture of sensory experience and wise understanding, a mixture that can set us down the path toward a more balanced way of being human.

FURTHER READING

Lee, K. E., Williams, K. J., Sargent, L. D., Williams, N. S., & Johnson, K. A. (2015). 40-second green roof views sustain attention: The role of micro-breaks in attention restoration. *Journal of Environmental Psychology, 42,* 182–189.

Mooney, C. (2015, May 26). Just looking at nature can help your brain work better, study finds. *Washington Post.* Retrieved from https://www.washingtonpost.com/news/energy-environment/wp/2015/05/26/viewing-nature-can-help-your-brain-work-better-study-finds/

Raskin, A. (2011, January 4). Think you're good at multitasking? Take these tests. *Fast Company.* Retrieved from https://www.fastcodesign.com/1662976/think-youre-good-at-multitasking-take-these-tests

Gloria, M., Iqbal, S. T., Czerwinski, M., Johns, P., & Sano, A. (2016). Neurotics can't focus: An in situ study of online multitasking in the workplace. In *Proceedings of the 2016 CHI Conference on Human Factors in Computing Systems.* New York, NY: ACM.

Killingsworth, M. A., & Gilbert, D. T. (2010). A wandering mind is an unhappy mind. *Science, 330*(6006), 932–932.

Robinson, K. (2015, May 8). Why schools need to bring back shop class. *Time.* Retrieved from http://time.com/3849501/why-schools-need-to-bring-back-shop-class/

TINYpulse. (2015). *2015 Best Industry Ranking. Employee Engagement & Satisfaction Across Industries.*

INDEX

ABOUT THE AUTHORS

DANIEL Z. LIEBERMAN, MD, is professor and vice chair for clinical affairs in the Department of Psychiatry and Behavioral Sciences at George Washington University. Dr. Lieberman is a Distinguished Fellow of the American Psychiatric Association, a recipient of the Caron Foundation Research Award, and has published over 50 scientific reports on behavioral science. He has provided insight on psychiatric issues for the US Department of Health and Human Services, the US Department of Commerce, and the Office of Drug and Alcohol Policy, and has discussed mental health in interviews on CNN, C-SPAN, and PBS. Dr. Lieberman studied the Great Books at St. John's College. He received his medical degree and completed his psychiatric training at New York University.

Trained as a physicist, **MICHAEL E. LONG** is an award-winning speechwriter, screenwriter, and playwright. As a playwright, more than 20 of his shows have been produced, most on New York stages. As a screenwriter, his honors include finalist for the grand prize in screenwriting at the Slamdance Film Festival. As a speechwriter, Mr. Long has written for members of Congress, US cabinet secretaries, governors, diplomats, business executives, and presidential candidates. A popular speaker and educator, Mr. Long has addressed audiences around the world, including in a keynote at Oxford University. He teaches writing at Georgetown University, where he is a former director of writing. Mr. Long pursued undergraduate studies at Murray State University and graduate studies at Vanderbilt University.